Wild North Carolina

T0248249

Wild North Carolina

DISCOVERING THE WONDERS OF OUR STATE'S NATURAL COMMUNITIES

David Blevins & Michael P. Schafale

THE UNIVERSITY OF NORTH CAROLINA PRESS   CHAPEL HILL

This book was published with the assistance of the Blythe Family Fund of the University of North Carolina Press.

© 2011 THE UNIVERSITY OF NORTH CAROLINA PRESS

Photographs © 2011 David Blevins
All rights reserved. Designed and set by Kimberly Bryant in Warnock Pro with Dear Sarah Pro and Avenir display. Manufactured in Canada.

The paper in this book meets the guidelines for permanence and durability of the Committee on Production Guidelines for Book Longevity of the Council on Library Resources.

Library of Congress Cataloging-in-Publication Data
Blevins, David, 1967–
Wild North Carolina : discovering the wonders of our state's natural communities / David Blevins and Michael P. Schafale.—1st ed.
    p. cm.
Includes bibliographical references and index.
ISBN 978-0-8078-3467-1 (cloth : alk. paper) | ISBN 978-1-4696-8335-5 (pbk. : alk. paper) | ISBN 978-0-8078-7779-1 (ebook)
1. Natural history—North Carolina. 2. Natural areas—North Carolina. 3. Biotic communities—North Carolina. I. Schafale, Michael P. II. Title.
QH105.N8B56 2011
508.756—dc22   2010034638

Contents

Authors' Note

Author and ecologist Aldo Leopold once said, "One of the penalties of an ecological education is that one lives alone in a world of wounds. Much of the damage inflicted on land is quite invisible to laymen." We have both had to face this feeling, David while completing a Ph.D. in forest ecology and Mike in his years of work with the North Carolina Natural Heritage Program. At the same time, recognizing these wounds allows us to better value the healthy places where nature remains free to express itself. Producing this book allowed us to focus on the love and wonder we find in nature that motivated us to become ecologists in the first place. As you take this journey to discover wild North Carolina, we hope you will share these feelings with us.

This book is the product of a fortunate coincidence. Both of us began thinking of a book like this before we met, but once we found each other our collaboration allowed us each to focus on our strengths. The authors' names are listed alphabetically, in keeping with the publishing tradition for equal contributors. Although Mike wrote the text and David made the photographs, both text and photographs were created through a dialog between us. We do not view this book as belonging to us, as much as we belong to it. Through this book, we offer our appreciation of natural communities and our love for wild North Carolina.

Introduction to Natural Communities

If you have hiked through North Carolina's parks, driven the Blue Ridge Parkway or another of the state's scenic drives, or paddled its rivers, you have seen natural communities. Perhaps you have wondered about some of the patterns—why you can hike for miles under oak trees and suddenly find yourself amid beeches. Or how you can climb a mountain slope and go from open, fern-filled forest to a hedgelike wall of evergreen bushes. Even if you can't identify the kinds of trees you encounter, you might notice that the trees with the warty bark are always near rivers along with the trees with the blotchy, pale brown, smooth bark. Perhaps you have noticed that the green of the mountains laid out below an overlook is a slightly different shade on the ridges than between them.

These patterns, if you learn to recognize them, can help you better appreciate and make sense of the complexity of the natural landscape. Understanding them helps transform what at first seems like a tangled green wall of vegetation into a meaningful picture of the natural world. Paying attention to natural communities will help you see familiar places in a new way and new places with a sense of familiarity. Look for them, and the details that continue to unfold can feed your curiosity for a lifetime.

WHY CARE ABOUT NATURAL COMMUNITIES?

Natural communities work for you, whether you are aware of them or not. They carry on the processes that keep your world inhabitable: producing oxygen, recycling nutrients, building soil, tempering floods, and filtering waste. They are the home of countless wild animals and plants, giving them food, shelter, and a place to raise their young. In the diversity of their living organisms lies not only the resilience of the ecosystems we depend on, but the future ecosystems that will develop as climate changes and continents shift.

If you come to appreciate them, natural communities will reward you in other ways too.

It is easy to see the repeating patterns formed by natural communities on the landscape in the Roan Highlands. Open grass balds on the knobs give way to northern hardwood forests on the slopes below. The higher peak beyond is covered in darker spruce-fir forests. Mountain oak forests cover the lower ridges in the distance.

In them, you may find mental and spiritual respite. They will offer you new insight into that world of nature that exists beyond human creation. And they will show you beauty you might have missed.

WHAT IS A NATURAL COMMUNITY?

The North Carolina Natural Heritage Program defines a natural community as "a distinct and recurring assemblage of plants, animals, fungi, and bacteria, in association with each other and with their physical environment." Like human communities, natural communities are made up of many different members, all sharing something in common even as each has its own personality. As in human communities, the different members of a natural community are there for their own reasons—the environment of a particular place makes it possible for some species to live there, while preventing other species from doing the same. Those species of plants, animals, and other organisms that thrive in the same kind of environments tend to show up together. But they also interact, helping each other or competing for space. Interactions make it possible for some species to live together, while driving others out. Though a few members may dominate, setting the terms for others' existence, many members make up the community. Communities differ from place to place. But go to a similar environment and you are likely to see similar communities. They won't be exactly the same, but they should share enough that you can recognize a pattern—a recurring assemblage.

When you look at natural communities, it generally works best to focus on the vegetation and on the physical environment. While the animals, fungi, and bacteria are crucial members of the community, the plants make up most of what you see. If you know the vegetation, it is generally easier to predict the animals than the other way around. Most of the bacteria and fungi, even the insects and other small creatures that make up most of the animal component, are too small to see without special equipment. Even the larger animals are good at hiding. The easily visible animals—deer, squirrels, most of the birds—tend to be those that range widely through many different kinds of communities, and therefore tell you the least about the community you are in.

LOOKING AT VEGETATION

A change from one natural community to another may be obvious when you come out of the woods onto a barren rock cliff or grassy dune. But many of our natural communities at first glance look alike, forming a nondescript green backdrop of trees. Paying attention to the species of plants and the shape of the land can help you recognize the more subtle differences in communities.

How do you look at vegetation? A good place to start is structure. What general kind of plant dominates: trees, shrubs, herbaceous (nonwoody) plants? Or is the vegetation so sparse that there isn't really a dominant kind of plant? Do the tallest plants form a dense canopy, as in a forest with nearly continuous trees blocking out the sky? Is the canopy open, as in a savanna with well-spaced trees? Or are the tallest plants grasses or shrubs, leaving you an open view of the sky? The diversity of species is another important feature of a natural community. Some communities are very diverse, while others are dominated by just a few species. Does the canopy consist of one kind of tree, a few, or many? Are there a few main species of herbaceous plants, or are there many? The types of plants that make up a natural community are also important. Is the forest canopy dominated by oak trees, pine trees, other kinds? Do the shrubs mostly have thick, evergreen leaves or thin, deciduous leaves? Are the herbaceous plants mainly grass, ferns, or broadleaf wildflowers?

LOOKING AT THE LAND

The physical environment, while made up of many different characteristics and forces, includes some that are easily observed and helpful in determining why one kind of natural community rather than another occupies a given place. Of the important aspects of the environment, one of the easiest to see is topography. Are you on top of a ridge, on a slope, or in the bottom of a valley? How big a landform is it? Is the ridge top broad or narrow? Is the slope steep or gentle? The slope angle influences how fast rainfall runs off and, therefore, how dry or moist a community is. What direction does the slope face? Does the sun shine on it all day, or is it shaded by a hill for part of the day? This affects how cool or warm the environment is, which also affects moisture. How high above sea level are you? Higher elevations are colder and more moist. Are there smaller landforms within the community? Are there hummocks and hollows, or low ridges and swales? Even very small surface variations can affect moisture; they can be especially important in wetlands. What can you tell about moisture levels? Does this place seem more or less moist than other places nearby? Are your feet wet? Is it too damp to want to sit on the ground, despite no rain for a couple days? If it is wet, does the water flow or stand? How deep? Are there water lines on the tree bases, or stains on the lower plants? Does the water come from river floods, from tides, from seepage out of the ground, or only from rainfall? Are there piles of debris moved around by flowing water? What about the soil? Is it rocky? Sandy? Is it soft underfoot? Is there even soil at all, or is it mostly bedrock? What

A narrow piedmont floodplain forest of sycamore and river birch lines the river in Eno River State Park. Behind the floodplain on the right is a moist slope with beech trees.

else might go on in this place? Is there charcoal on the ground or at the bases of trees, to suggest that it has burned recently? Are trees or shrubs gnarled or streamlined, suggesting wind has shaped them? Are there beaver dams, active or abandoned?

Some important aspects of the physical environment are not easy to see. Water may be present at other times and leave no sign. Soils may be fertile or infertile, but only a lab test will tell for sure. Once you start learning patterns, you can often infer some of these characteristics as well. Even if you don't know many specific kinds of plants, you can make some educated guesses. Small leaves, thick leaves, and hairy leaves help plants conserve moisture. If you see these characteristics in most of the plants in a place, it likely is dry. In contrast, if you see mostly plants with broad, thin leaves, the place is likely moist. Many species of plants grow

Introduction to Natural Communities

only in specific environments. If you see plants that you've seen only in wet conditions before, and none that you've seen in obviously dry sites, you can conclude the place is probably usually wet even when you see it bone dry. Though harder to sort out, the same is true for plants that, for example, require particularly fertile soils, or that tolerate particularly infertile soils, or that tolerate salt.

LOOKING AT LAND HISTORY

Natural communities are products of nature, the result of the physical environment and the interactions of different living beings. Natural communities once covered the landscape throughout North Carolina, but after centuries of intense land use, they have become hard to find in many regions. Even many forested places that appear unused at first glance do not resemble what nature would put there on its own, and will not match anything we describe in this book. This is true of most places that once were plowed and of many forests that were clear-cut in the last few decades. Thus, one of the first questions to ask, and sometimes one of the hardest to answer, is whether a place is a natural community at all. Good examples of most kinds of natural communities are now rare, and it often takes a special effort to find them.

Any homeowner with a lawn or garden knows that, without ongoing human intervention, vegetation will grow back on cleared land. Plants that are good at spreading their seeds far and wide, from creeping crabgrass to shaggy broomsedge to towering loblolly pine, will appear on disturbed land. Left to their own devices, they often come and go in a predictable order, with larger plants supplanting smaller ones over the years. This kind of growth is known as successional vegetation. The widespread loblolly pine forests of central and eastern North Carolina, the sweet-gum and red maple woods in eastern wetlands, and the tulip poplar and black locust forests of mountain valleys are all examples of successional vegetation that has grown up following the destruction of natural communities.

Eventually, more and more members of the natural community may find their way back to replace the successional vegetation. However, as with a person who has survived a severe injury, the community may not be the same afterward, or it may take a very long time to recover. Most farmland abandoned in the chaos of the Civil War or the Great Depression of the 1930s still supports successional vegetation rather than recovered natural communities.

Forests that were selectively logged or that were clear-cut longer ago may be recognizable as natural communities but will be different from less disturbed examples in subtle ways. After centuries of human land use, even the most pristine places have seen some impacts. How do you look for evidence of natural character or loss of it? Can you see the faint parallel ridges that show that the soil was once plowed or that it was bedded for planting pines? Are there sawed tree stumps? Are there ruts from old roads or from logging equipment? Are there ditches that might have drained water away? Many of these kinds of alterations can also be inferred from the vegetation. Are the trees small and young-looking compared to those in other places, without downed trunks on the ground to indicate the effects of a recent storm? Are the trees all the same size, suggesting they all got their start at the same time? Are they mostly of species that tend to invade cleared areas—such as loblolly pine, tulip poplar, or black locust? There are also signs you can seek to indicate that a site has not had such a history and is relatively undisturbed. Are there any very large trees—three feet in diameter or more? Old forests aren't necessarily made up only of such trees, but they usually have many. Are there large, old logs on the ground, especially ones that clearly fell at different times because they are in different stages of decay? Are there remnants of the low mounds formed when trees uproot in storms? In old forests, such mounds persist long after the fallen tree has rotted away, with the hint of a pit next to them where the roots pulled out.

One additional thing you can do to help make sense of natural communities is to think about where they occur. Many aspects of the physical environment follow broad geographic regions. Some are obvious: the mountain region tends to have steep slopes and high elevations, while the area near the coast does not. Some are subtle, such as the near-total absence of bedrock and the much wider river floodplains in the eastern part of the state. Some are nearly invisible, such as the abundant sandy soils in the eastern part of the state and their near absence in the western part. Living organisms follow geographic patterns, too, with some species occurring only in one region, or at least being largely absent in another. So, often, knowing where you are helps tell you what species or natural community types are even possibilities. Regions also vary in what aspects of the physical environment are most important to look at for determining different natural-community types. Most types of natural communities occur in only one of the three regions of North Carolina. Some others occur mostly in one region but can be found in unusual parts of another. Only a few types are common in two or all three regions. The names of the types often indicate where they occur, and the descriptions in each chapter will help indicate where you should look for them.

North Carolina is traditionally broken into three major regions: coastal plain, piedmont, and mountains. These region names offer a quick way to refer to the many physical and biological differences between them. Since we will mention these regions in naming and discussing community types, a brief introduction is in order.

The mountains, really the Blue Ridge Mountains, are the smallest region, making up the western end of the state. While there are some sizeable valleys and basins, most of the land in the mountains is rugged and steep, and elevations are higher than in other parts of the state. Higher elevations bring cooler temperatures and more rainfall, but there are also major differences in rainfall within this region. The southern portion, along the South Carolina and Georgia border, has the highest rainfall in the eastern United States, while the valley around Asheville is fairly dry. In the mountain region, topography and elevation are the most important aspects of the physical environment. The boundary of the mountain region is the Blue Ridge escarpment, an especially steep slope that forms the blue wall you see when you

On the Outer Banks of North Carolina, you can easily see how natural communities are organized on the landscape. On wild beaches like this one at Cape Hatteras, natural communities change as you move from the dunes to the sandy flats and wet swales, and to the maritime forest in the distance.

approach the area from the east and that forms the perch for many of the overlooks on the Blue Ridge Parkway north of Asheville. However, some mountainlike foothills exist east of the escarpment, including the South Mountains, the Brushy Mountains, and the Sauratown Mountains.

The piedmont is the middle part of North Carolina, home to most of the cities and most of the people. It was part of the Appalachian Mountains at the height of their glory in the geologic past. A long period of erosion has reduced piedmont topography to a rolling landscape punctuated with higher remnant hills. River valleys and the dissected land along them are the main form of topography. Isolated remnant hills, known as monadnocks, are prominent in some places. The piedmont has a typical warm, southern climate and generous amounts of rainfall that are normally well distributed through the year. There is a greater diversity of rock types in the piedmont than in the other two regions, and their variation has important effects on natural communities. The type of rocks and soil, along with topography, are the most important aspects of the physical environment for natural communities in the piedmont.

It is not necessarily easy to tell when you have left the piedmont as you travel eastward or southward. That edge of the piedmont is marked by the Fall Zone, a band where the hard rocks dip beneath the soft sediments that make up the coastal plain. Although the coastal plain is generally lower in elevation than the piedmont, in the Fall Zone the coastal plain sediments lie on top of the hills, while the underlying piedmont rocks can be found along the rivers. The same zone is the location of the easternmost rapids on the rivers, an important boundary for human travel in the early days of European settlement.

The coastal plain is the largest region of North Carolina, forming the eastern side of the state and making up almost half its area. Like the piedmont, its climate is warm and moist. The coastal plain is geologically young. It consists mostly of sand and clay sediments that never hardened into rock. All of it was submerged by the ocean in the recent geologic past, with parts near the coast having emerged just thousands, rather than millions, of years ago. The coastal plain is largely flat, but there is subtle topography created by the meandering of rivers across the level landscape, or by the waves, tides, and blowing sand of past coastlines. These affect soil fertility and wetness, and natural communities usually follow their patterns closely. Water is particularly important in the natural communities of the coastal plain, and this region demonstrates that there are many different ways of being wet. How much water, where it comes from, how long it stays, and how salty it is varies widely and is the major reason why the coastal plain has more types of natural communities than the other regions. The coastal plain is also the part of the state most prone to fire, thanks to an abundance of thunderstorms and a scarcity of hills and slopes to act as natural firebreaks. In addition, much of the vegetation itself is more flammable than in other regions. Though it is difficult for most people to picture now, fire was an important natural force in all of the state. It mostly disappeared from the landscape only a couple of generations ago, and its imprint on natural communities is still very important. Many of the natural communities in all three regions were shaped by fire, but this is particularly true in the coastal plain.

Two particularly distinctive parts of the coastal plain bear special mention. The tidewater region is the outermost part of the coastal plain, where streams are at sea level and the land is barely higher. The influence of the ocean, in the form of salt water and tides, is important here. Because the land is so low and flat, rainwater runs off slowly, and most places are wet. The sandhills region, located along the inland edge of the coastal plain in southern North Carolina and farther south, is an area covered with a broad expanse of gently rolling ancient sand dunes. Because this area is fairly far inland and fairly old, streams have dissected the landscape more than in the rest of the coastal plain, almost as much as in the piedmont. The result is a hilly landscape with distinctive sandy soils, many slopes, and numerous small stream valleys. In all parts of the coastal plain, there are deep sandy soils that let rainwater drain through rapidly, creating dry conditions for natural communities. This kind of soil predominates in the sandhills.

CLASSIFICATION OF NATURAL COMMUNITIES

If you learn to observe the visible characteristics of the natural communities you encounter, you will enjoy seeing many details of the world that you otherwise might have missed. But how do you make sense of all these details you may observe? Though it has a long way to go, scientific study has recognized many of the patterns and deciphered their ecological importance. Classifications have been developed to help bring orderly understanding out of complexity, to help figure out which variation is important and which things are not as different as they might seem.

This book uses a classification system developed by scientists at the North Carolina Natural Heritage Program to sort out the different kinds of natural communities. The most detailed classification, used for technical purposes by the Natural Heritage Program, describes several hundred different types of natural communities. The program's more general classification, which we use here with some modification, covers the major variety of natural communities and gives a good framework for appreciating that diversity. Though the primary purpose of this book is not to teach you to recognize all the different types of natural communities, each chapter will describe important aspects of what distinguishes the types. (We use common, rather than scientific, names for plants and animals throughout this book, but the scientific names are given parenthetically in the index entry for each.) We will also offer a bit of what is known about the ecology of each, which may help you better understand and appreciate the wonders of our natural communities.

Spruce-Fir Forests

In North Carolina and its neighboring states, spruce-fir forests crown the highest mountains. From a distance, you can see their dark green color on the peaks and ridges, contrasting with the yellower shade of green or the winter gray of the hardwood forests below. If you are fortunate enough to walk in a spruce-fir forest that is close to its primeval form, it is an experience like no other in our state: cool, shady, damp, green from top to bottom, and perfumed with the scent of Christmas. Even the tree trunks can be green, covered with mosses and liverworts, while the ground may be covered with an inviting bed of moss, feathery ferns, or a lush but delicate carpet of the cloverlike leaves of wood sorrel. If the restless mountaintop wind is quiet enough, the sound of water dripping off trees and trickling between rocks underfoot breaks the wilderness silence. The chatter of a red squirrel, the croak of a raven, or the chirps of more secretive birds may be the only other sounds. If there is a break in the canopy, you may be able to see out over the neighboring mountains. Perhaps just as likely, you may find yourself in a secret enclosure of overarching trees and fog, with only a vague feeling of space beyond to tell you that you are at the top of North Carolina.

This pleasant experience at the height of summer can be a welcome relief from the sweltering heat of the lowlands. Visit a spruce-fir forest at other times of the year, or in other weather, and you may experience something quite different. Unrelenting wind may blast across the ridges; lightning may crack on the peaks; plummeting temperatures and drenching rain may turn a casual visit into an emergency for the unprepared. In the heart of winter, the coolness translates into some of the most severe weather North Carolina experiences. Besides the familiar blanket of snow, you may see rime ice—hoarfrost formed from fog droplets freezing onto trees, rocks, even buildings. Rime ice may be a thin coating, or it may grow into long needles of ice that stick out like shaggy fur, looking windswept but actually growing into the face of the wind.

Spruce-fir forests owe their existence to the cool, wet climate that accompanies high elevations throughout the world. Rising air cools, and cooling air concentrates whatever moisture it

(opposite)
Huge red spruce emerge from feathery ferns into a damp summer fog.

(overleaf)
Fraser fir trees struggle to survive introduced insects, acid rain, and a harsh environment. These ghostly trees at Mount Mitchell State Park have succumbed, but young fir saplings beneath them may be able to form a new forest canopy in time.

carries. The result is plenty of rain and fog. The high peaks get up to seventy or eighty inches of rain a year, nearly double the rainfall in the nearby mountain valleys. And no place is better suited for showing that fog really is made of water. The shrouded forest drips, as spruce and fir needles strain the fog droplets from the air. This dripping fog may give the mountaintops as much moisture as they get from rain and snow.

While abundant moisture is generally favorable to plants, the cold winters and short summers limit what plants and animals can occur on the high mountains. Few species are shared with the lower slopes of the same mountains. Though spruce-fir forests may have dense vegetation from the tree canopy down to the herbaceous ground cover, they have a limited variety of species. While you may see more kinds of trees than just spruce and fir, you won't see many. Yellow birch, with its silvery, striped bark, is common. Mountain ash, our feathery-leaved relative of the European rowan, may also be abundant. Small maples may occur in the understory. These largely exhaust the list of trees. And a look at the lush herbaceous layer may reveal just one or two species: Canada mayflower, wood sorrel, whorled aster, or one of several types of ferns. Even in this extreme environment, though, with few species able to survive the climate, competition among plants is important. The dense shade of the canopy keeps out some species that would otherwise be able to grow on the mountaintops.

Our spruce-fir forests are sometimes called "boreal" or "Canadian" forests. And, indeed, if you visit most parts of eastern Canada, you will see forests of spruce and fir, with many familiar plants and animals. This is to be expected, since climate tends to trade off altitude for latitude; climbing higher in elevation results in colder average temperatures, just as going north does. The connection of our spruce-fir forests to Canada is only a general one though, and the plants and animals illustrate this well. Our spruce species, red spruce, extends throughout the Appalachian Mountains, up into New England, but is not the common spruce of Canada. Our fir species, Fraser fir, is found only in the southern mountains, barely extending beyond North Carolina's borders. Witch's-hobble, the round-leaved shrub whose ghostly white flowers light the shady forest, is widespread in the North; while mountain cranberry, another spruce-fir forest shrub, gets no farther north than West Virginia. The red squirrels that chatter in the trees are common in the North, while the Appalachian northern flying squirrel is actually a rare subspecies distinct from those of the North. Weller's salamander is found on only a few high peaks in North Carolina, Tennessee, and Virginia. Entire species of some ground beetles, meanwhile, may be confined to a single North Carolina peak.

If you have visited one of North Carolina's high mountains recently, you may be wondering why you didn't see the primeval-looking forest described here. Instead, you may have seen a ghost forest of dead trunks with a thicket of Christmas trees beneath, or a tangle of blackberry briers. From a distant ridge, the peaks may have been the white of dead wood rather than a dark, evergreen green. The characteristic canopied green world of old spruce-fir forests has become scarce. While much of North Carolina's original spruce-fir forest was destroyed by the sawyers and steam skidders of the early 1900s, most of what survived became protected in national parks, national forests, state parks, and private conservation lands. Spruce-fir forests are now less subject to deliberate destruction than any other forest type. Yet, somehow, because of their unique character and habitat, along with sheer bad luck, ecological disaster has overtaken them anyway. Despite their remoteness, the mountaintops tend to concentrate air pollution from the wider surrounding area. Ozone, heavy metals, and acid rain are all suspects in the damage. It turns out the fog is even more acid than the rain, making these foggy mountaintops particularly susceptible to its effects.

Weller's salamanders are found on only a few high peaks in North Carolina, Tennessee, and Virginia.

The worst culprit, however, is a small aphidlike insect accidentally brought from Eurasia, the balsam woolly adelgid, sometimes noticeable as tiny, fuzzy white dots on the trunks of fir trees. This insect sucks the sap from many species of fir, but Fraser fir has proven susceptible not only to the bleeding but to something more like a lethal allergic reaction. The tree's water-carrying cells react to piercing by the adelgids by thickening their cell walls, closing off water transport. Starting in the late 1950s, balsam woolly adelgids appeared in one mountain range after another. While seedlings and saplings were not affected, within a few years all mature Fraser firs were dead. At the highest elevations, where fir was more common than spruce, the result was a ghost forest of snags. Even where the trees were more evenly mixed, the death of half of the trees broke up the shady canopy, allowing in blackberries and other plants that need abundant light to proliferate. It also let in drying sunlight and winds, to the detriment of the mosses and liverworts. Only the forests at lower elevations, where spruce prevails and fir was a minor component, may retain their primeval look.

A visit to the spruce-fir forests now offers a glimpse of the race between destruction and hope. Many of the ghost forests have now fallen. Shrub thickets and blackberry patches are well established in many higher areas. But spruces and birches have filled in the empty spaces in lower-elevation forests. Having killed off most of their food source, balsam woolly adelgids are now much scarcer than they were. With the young firs not susceptible to the adelgids, a new generation of firs has quietly been growing up from the seedlings that waited beneath the dead canopy. They now are again forming a forest canopy in some places, though they are far from their former majesty. As the firs grow, they become susceptible to balsam woolly adelgids, and some are dying. Will they reach a size where they can produce seeds and get a crop of seedlings established before they succumb? If they succeed, Fraser fir may remain a part of these forests, though perhaps in smaller numbers, perhaps living thirty years instead of eighty. If they remain and are able to keep reproducing and trying new genetic combinations, there is hope that they will eventually evolve greater resistance to the insect that now threatens their survival as a wild species.

The extent of spruce-fir forests is very limited compared to most of our forests, but they have received more attention from both scientists and the public than their scarcity would suggest. It was the attraction of the spruce-fir forests and the destruction of logging them that led to the creation of North Carolina's first state park, Mount Mitchell, in 1917. The high tops of the Black Mountains had already been an attraction to visitors for decades by then, for their beauty and distinctive feel.

Though spruce-fir forests occur on remote mountaintops, they are no longer hard to visit. Some are readily accessible in Great Smoky Mountains National Park, along the Clingmans Dome Road and at Clingmans Dome itself. Spruce-fir forests can easily be seen at Mount Mitchell State Park, Grandfather Mountain, Roan Mountain, and in scattered patches along the Blue Ridge Parkway around Richland Balsam. If you are a hardy day hiker or backpacker, you can find extensive spruce-fir forests, including old-growth forests, in the remote backcountry in the eastern half of Great Smoky Mountains National Park.

The Carolina dark-eyed junco breeds in the high mountains of the southern Appalachians and does not migrate like the more commonly seen juncos from farther north. Carolina juncos breed on the tops of high mountains and migrate downslope for winter. The juncos that breed to the north migrate long distances and are commonly found across North Carolina in winter. Carolina juncos are the only juncos in the state in the summer; in the winter you can distinguish them from their northern relatives by their stouter white bill.

Northern Hardwood Forests

The northern hardwood forests lie high on the mountains, where they are usually the highest of the hardwood forests. They crown the ridges and peaks in many ranges, and run up to blend with the spruce-fir forests in the highest ranges. You can pick them out easily from a distance in the early fall, when they are the first to show the change of color. If you have wandered the glowing orange mountain wonderland of mid-October, your most memorable moments of brilliance were probably in the northern hardwood forest. In summer, they may seem just another part of the endless green forest, but walk into one and you may notice that oak leaves have given way to the oval leaves of birch or beech, or perhaps to maple. The trees may be as tall and majestic as in any other mature forest, but climb to a high knob or a windswept gap and you may find gnarled and stunted veterans of many an icy blast.

In the warm summer, northern hardwood forests can be some of the most inviting natural communities. With little understory, the sunlight dapples through the tree canopy to the shrubs and herbs below. You can lie down on a lawn of fine grasslike sedges, or hike over long slopes covered with lush herbaceous growth. The bird chorus of early summer may reward you with the long, lilting song of the winter wren or the strange circling song of the veery. The low, accelerating thumps of a grouse's drumming may rise from the slopes below. In spring, a dense carpet of small wildflowers often carpets the ground: spring beauties, trout lilies, lacy-leaved squirrel corn, pungent ramps. As in other high-mountain communities though, you may find your pleasant stay cut short by chill, fog, or storm even in the warmest season.

Northern hardwood forests are tied to the cool, moist climate of high elevations, though they give way to spruce-fir forests in the rigors of the highest elevations. Here, plants and animals that occur in little moist pockets at lower elevations spread out to cover the ridges and open slopes. The lower temperatures that come with high altitude give plants less need of water here, just as your garden needs less watering in cool weather. Even this high, though, other aspects of topography play a role in creating a suitable home. You most often see northern hardwood forests on slopes that face north or east. Cross a ridge to a west-facing side,

Stunted American beech cover a lawn of fine sedges along the Blue Ridge Parkway.

where the sun shines more strongly in the warmer afternoon, and you can find yourself back in oak forest.

Northern hardwood forest natural communities are made up of more plant and animal species than are spruce-fir forests, but the number of species tends to be lower than in moist communities at lower elevations. Though you may sometimes see six, even eight kinds of trees in the canopy, more often it is four; and sometimes it is nothing but yellow birch or beech over many acres. While most of the plants of the northern hardwood forests also occur at lower elevations, most plants of lower elevations are excluded by the short growing season. The animals are also less varied than at lower elevations but include some unusual species such as the secretive and rare Appalachian northern flying squirrel.

Northern hardwood forests are named for their similarity to the moist hardwood forests of the northern and upper midwestern United States. However, as with the spruce-fir forests, this similarity is only relative. A visiting botanist or zoologist from New England would see many old friends in these communities, but perhaps as many strangers. Despite the name, our northern hardwood forests are distinctive southern communities, occurring only in North Carolina and its neighboring states.

One distinctive form of northern hardwood forest is worth a special mention. If you hike the high mountains enough, you may come across an area where the ground is completely covered with large boulders, so thickly that virtually no soil has been able to accumulate at the surface. During the ice ages, freezing and thawing fractured rocks and caused them to creep downhill and pile up in some high coves. While the Rocky Mountains and the northern Appalachians have boulder fields that are often devoid of vegetation, ours are not barren. The rocks are covered with moss and ferns, maybe low gooseberry bushes, and a canopy of trees shades the ground almost as densely as on the deep soils nearby. Yellow birch is particularly well suited to living in these boulder fields, because its seedlings can germinate and survive on rocks or logs. Even in forests with deep soil, you often see some yellow birch trees standing on stiltlike roots that once wrapped around a now-vanished log. In the boulder fields, wrapping roots around obstacles to reach the soil below is the only way for a tree to live, and yellow birch is often left

Squirrel corn is a common early spring flower in northern hardwood forests and mountain cove forests.

as the sole possessor of the canopy. The honeycomb of spaces between the boulders makes for dangerous walking for people but offers especially good hiding places for small mammals such as rodents and shrews. Though you are unlikely to see them, a surprising number of these animals likely lurk beneath your feet in the boulder-field forests.

The ice ages had other, less obvious, effects on our mountains, ones that the northern hardwood forests make an especially good place to contemplate. In several mountain ranges, such as the Craggy Mountains north of Asheville, the highest peaks reach elevations that normally

support spruce-fir forests, but no spruce-fir forest is present. Instead, northern hardwood forests cover the peaks, reaching higher elevations than they do in other ranges, such as the nearby Black Mountains. Why should this be, when mountain vegetation usually reflects elevation more precisely?

As the climate has cooled and warmed through the ages, plants and animals have had to move to adjust. When it was colder, northern hardwood forests and spruce-fir forests crept down the slopes as the communities of lower slopes were driven down into the warmer valleys. About 6,000 years ago was a period when the opposite occurred, when the climate was warmer and drier than at present. Then, heat and drought and fire drove the northern hardwood forests and spruce-fir forests even higher up the mountains than we now see them, so that they could survive only on the highest peaks. In ranges as high as the Craggy Mountains but no higher, it appears that the spruce-fir forests found no place cool enough to survive that warm period. They were "pushed off" the top of the mountains. With most of the plants and animals of the spruce-fir

Painted trillium is one of the typical spring wildflowers of northern hardwood forests.

forests unable to cross the lower lands to return to these ranges, those of the northern hardwood forests had no competition for the high peaks. You might notice, however, that northern hardwood forests at the top of these ranges are particularly stunted, and that they contain some of the plants of spruce-fir forests. These movements of communities offer a picture of what we likely will see as the world's climate warms in our own time. It illustrates the resilience and ability to adjust that natural communities have, but also the loss that comes when they are pushed too far.

Northern hardwood forests can be found in all of the higher mountains of North Carolina. You can drive through them or hike through them on good trails along the Blue Ridge Parkway and in the adjacent national forests in the Craggy and Black Mountains and in the ranges south of Asheville. Great Smoky Mountains National Park has substantial areas of old-growth northern hardwood forest along hiking trails, as well as examples visible from roads. Grandfather Mountain has good examples reachable along its rugged trails or visible from the Blue Ridge Parkway. Examples of places where northern hardwood forests extend up to elevations that could support spruce-fir forest include the Craggy Mountains and, farther north, Elk Knob State Natural Area.

(opposite)
Yellow birch and American beech stand amid ferns on a foggy autumn day in the Craggy Mountains.

Grass and Heath Balds

In a mountain region dominated by dense forests, the open expanses of grass and heath balds are some of the most distinctive natural communities. Look at the right ridge and you may see the smooth green "slicks" of heath balds or the tawny color of grass balds standing out against the darker green or winter gray forests around them. Some of these balds cover domes or broad ridge tops, as at Roan Mountain and Craggy Gardens, while others drape narrow, plunging ridges, as many in the Great Smoky Mountains do.

While balds lack trees, they are not bare ground. In fact, they have dense vegetation. Up close, the grass balds are lush meadows, offering a view as sublime as any scenery North Carolina can provide. The lacy panicles of grasses and drooping spikes of sedges wave in the breeze. Colorful wildflowers punctuate the meadows: the big red-orange bells of lilies, the yellow of goldenrod, or the white of the creeping, strawberrylike three-toothed cinquefoil. The slopes fall away to the deep forests below and open outward to the rolling sea of forested mountains beyond. You can wander over the meadow with the world at your feet, or lie down in the grass amid the silence and gaze at the unfettered sky. On a calm, sunny summer day, you will think no place could be so idyllic. You do well not to go unprepared though, as even in summer a blasting wind, a rolling bank of chilly fog, or a pounding thunderstorm can come swiftly down upon you. Rough weather can appear surprisingly quickly, despite the warning such an expansive view would seem to afford. Heath balds can be a different sort of experience. If you are lucky enough to get above them they may appear as seas of green leaves, which are usually the deep green ovals of evergreen rhododendrons or mountain laurel. And in the early summer they may blaze with the pink of rhododendron flowers. But since the shrubs are usually taller than a person, you typically see them from below, as a phalanx of branches and trunks twisting and tangling beneath the leafy canopy. If there is a trail, it runs as a tunnel through the thicket. If there is no trail, the prospect of crossing a heath bald can be quite daunting.

Balds are an interesting ecological mystery. While we know enough about most of our natural communities to say what makes them what they are, or why they occur where they do,

(opposite)
Lush, windswept grasses and sedges dominate this view of the grass-bald community on Round Bald in the Roan Highlands.

this is not the case for grass and heath balds. In a region where none of the mountains are tall enough to reach timberline and the highest peaks are forested, they are notable. Why are they bald? The obvious cause in the rock-outcrop communities we will talk about next—lack of soil deep enough to support trees—is not the answer here. Why should these particular places be treeless? To be sure, they occur in rough neighborhoods, with intense cold and wind, but not more so than the spruce- and fir-covered peaks standing above them. The sharp ridge tops with shallow soils where some heath balds occur may be a surprisingly harsh setting for trees, but the locations of many heath balds and most grass balds seem indistinguishable from other sites that are forested.

This uncertainty doesn't come from lack of attention and thought. Though balds are rare, they have attracted plenty of scientific consideration and possible explanations. One theory is that they may not really be natural communities at all but may instead be results of clearing by early settlers or prehistoric Native Americans. Another is that large animals created them by tearing down and trampling out trees and then kept them open. The collection of large mammals that lived before and during the ice ages, known as the Pleistocene megafauna, included mammoths, mastodons, and less-talked-about creatures such as giant ground sloths, all capable of pulling down trees. After them, herds of elk and bison roamed North Carolina until the time of early European settlement, and herds of cows were put on these natural pastures soon after. Fire is also suggested as a cause. There are potential problems with all of these hypotheses. Some balds were already reported open at the time of earliest European settlement; and though Native Americans used them, they weren't major settlement sites. Some balds support rare plants of open habitats, which suggests they are ancient. And indeed, new openings created by logging or fire in forests typically remain open only a few years and don't resemble balds. The moist high mountains aren't very prone to fire in any case. And the Pleistocene megafauna have been extinct for some 10,000 years; their effect everywhere else faded long ago. There are potential answers to some of these problems, but none is proven. "None of the above" remains a viable choice for the origin of balds.

Besides the mystery of their baldness, these natural communities offer a paradoxical mystery: they don't stay bald. If you visit one of these communities, you may see that it is dotted with small trees or, in the grass balds, clumps of shrubs. Grass balds that appear to have been open since prehistoric times, probably for millennia, are being

The Gray's lily is a beautiful, rare wildflower found on grassy balds.

(opposite)
Blueberry leaves in a heath bald on
Grandfather Mountain turn a brilliant
red in the fall.

invaded by either trees or shrubs or both, some so rapidly that they are in danger of completely disappearing within our lifetimes. In fact, the only grass balds that have their characteristic treeless form anymore are those where it is maintained by active cutting, mowing, or grazing. Some heath balds appear to be stable, but others are turning into forests before our eyes. Whatever created and maintained the balds seems no longer to be creating or maintaining them. If that cause was Native Americans that have been gone for centuries, or large animals that have been gone for millennia, why would trees wait until our time to invade the balds? It is possible that grazing of livestock on the natural pastures of the grass balds, a practice that started with early European settlement and ended only in recent times, kept the trees out after the elk and bison were gone. It is equally possible that heavy grazing itself, or some other aspect of settlement, destroyed whatever mechanism had previously kept them open. In any case, the heath balds did not offer good grazing, yet many of them are also being lost. Thus, the gradual disappearance of balds remains a scientific mystery, just as their origin does. These mysteries have major implications for our natural community diversity and for the plants and animals that live in the balds. Compared to most of our natural communities, balds require a lot of effort to keep as part of our natural heritage even after the land is set aside.

While grass and heath balds occur on remote, high mountains, there are several places where they are easy to see. Craggy Gardens, on the Blue Ridge Parkway north of Asheville, and the Roan Mountain Gardens, on the Tennessee border farther north, offer views of heath

Flame azaleas add color to grass balds in June. While scattered shrubs are common in grass balds, excessive growth of shrubs threatens their open character.

balds on broad domes. You can see heath balds on sharp ridges along the Newfound Gap Road in Great Smoky Mountains National Park. The largest grass balds are on Roan Mountain. You can see them from the road at Carvers Gap and can hike through them on the Appalachian Trail. Other grass balds, Andrews Bald and Gregory Bald, make excellent day-hike destinations in Great Smoky Mountains National Park.

Grass and Heath Balds

High-Elevation Rock Outcrops

Like the grass and heath balds, rock outcrops are memorable places that provide variety amid the dense forests that cover most of the natural landscape. The forms of these outcrops are diverse. Some are jagged crags rising from the trees at the top of a peak; some are dizzying cliffs dropping off the side of a ridge. Some are smoothly curved domes of rock, whose gradually steepening slopes can tempt the unwary toward a dangerous slip. The openness of rock outcrops invites you to enter, or at least to sit at the top and enjoy the view of the plunging slopes and towering peaks, or of the vast, wrinkled green rug of the mountain expanses beyond.

With such views, it is easy to focus only on what lies in the distance. This is too bad, because if you do, you will miss a distinctive and fascinating natural community. Look closer at hand and you will see life at your feet. Even the crust of pale green or bright orange lichen, thin as a daub of paint on the bare rock, is alive. Moss, too, dense cushionlike clumps or forests of miniature pine trees, accomplishes the improbable feat of growing on solid rock. So, too, does spikemoss, which despite its name and appearance is more akin to ferns than to true mosses. Clumps of grass or sedges, easy to ignore but very much alive, sprout from the tiniest crevices. There may well be flowers too, diminutive white saxifrage, yellow dwarf-dandelion, or purple bluets, or larger and more showy ones, alone on the rock or arranged in little garden beds amid the austere expanse. Don't ignore the larger plants you stepped around either—the clump of blueberry bushes or rhododendron, the ragged birch tree or pine. And what animals might hide in those little thickets? Perhaps a chewed leaf or a pile of droppings will give you a hint.

Ignoring these small wonders is unfortunate for more reasons than just the beauty you miss. These are among the most fragile of our natural communities, one where the small plants are easily killed by just a few people walking on them. Many easily accessible rock outcrops have been trampled bare by visitors bent on views. Even the least accessible rock-outcrop communities have sometimes been seriously damaged by rock climbers.

High-elevation rock-outcrop communities and their inhabitants offer our most stark pic-

(opposite)
The Roan Mountain bluet, federally listed as endangered, is known from only twelve high-mountain locations in North Carolina and Tennessee, including this one on Grandfather Mountain. A small pocket of soil with minimal shelter allows this plant to thrive amid the bare rock. The battered shapes of the trees on the skyline behind attest to the harsh environment.

ture of the triumph of life over adversity. Like the spruce-fir and northern hardwood forests that surround them, they face our most extreme cold and shortest growing season, along with blasting wind, chilling fog, and rime ice. The trees that cling to them are often stunted and gnarled into bonsai forms reminiscent of those at the timberline of higher mountains. Added to this is the lack of soil. Thirsty roots are limited to the shallow mats that sit atop the rock, or to deeper but confining crevices. Plants grow where they can, a tree here, a patch of bushes there, a little bed of tiny flowers amid bare rock, without the distinct layers of a forest. Insects shelter amid the plants, and you may see a few salamanders. Hardy birds such as the raven and the very rare peregrine falcon occasionally nest on isolated cliffs. But most larger animals pass through, stopping to bask in the sun or perch on a branch but seeking shelter in the surrounding forest.

The plants that live in these scenic but difficult places are an interesting collection. Some never grow in soil, such as the crustose, paintlike lichens, the large, loose lichens known as rock tripe, and some of the mosses. Some, like the oatgrass and little bluestem, are equally at home in the grass balds or in forest openings below; while others, like the saxifrages, are shared with lower-elevation rock outcrops. Some are simply species of the surrounding forest, barely getting by where their seeds fell in some small pocket.

Perhaps the most interesting plants are those whose only home is on high-elevation rock outcrops. These communities are some of the best places in the mountains to find rare species. Some, like the colorfully named wretched sedge or the Blue Ridge goldenrod, stray into adjacent states, but just barely. Others, such as Heller's blazing star, with its purple spikes, grow only in North Carolina. The scarcest, like the mountain golden-heather, are among the rarest species in the world, clinging to existence only on the crags of one or two of North Carolina's mountains. As in the spruce-fir forests and other high-mountain natural communities, these narrow-ranging species are mixed with others that are rare in the South but widespread in the North. The Greenland sandwort, for example, ranges far into the Arctic.

Why should communities that are so hard to live in be home to so many rare plants, more than most of the lush forests and more than many communities that have greater overall diversity of species? There are several possible pieces to this puzzle, though proving any of them is hard. One is simple logic. High-elevation rock outcrops are so small and scarce that any species that occur only on them have no chance of being very common. But this leaves the question of why there are species that can live only there. So another part may be an ecological paradox that we will address more in later chapters—environments that are favorable enough that many species can live in them tend to be dense and crowded, with intense competition

(opposite)
Mats of spikemoss cling to the smooth surface of a high-elevation granitic dome in Panthertown Valley, providing a place for other plants to become established. Similar mats cling to the steeper face in the distance.

for light, water, and nutrients. Harsh environments have room for the smaller and less aggressive, as long as they can tolerate the other difficulties.

Another piece of the answer may lie in the ecological history of the region. We have already talked about the changes in climate as the ice ages came and went, and how the mountain communities shifted in response. Though there were no glaciers this far south, at the coldest times the highest peaks were too cold to support spruce-fir forests. Like the higher mountains of the Rockies, or the lower but more northerly mountains of New England, our mountains had a timberline then. And above the timberline lay a zone of alpine tundra—treeless vegetation of small plants that suffered the cold and wind but benefited from the bright sun. As the climate warmed and the fir and spruce spread uphill, it is likely that many of the hardy alpine plants hung on in the few cold, open habitats that remained—the high-elevation rock outcrops and grass balds. Thus, the distinctive rare plants of these communities would represent the last remnants of species that once were common, and of a community that no longer exists in North Carolina. As the world's current warming trend continues, more of our high-mountain plants may become rare species confined to the harsh refuge of the rock outcrops. But if they can survive, they may have a larger role to play in some future version of our unstable climate.

While the rigors of the mountaintop climate are generally more influential than the kind of rock in these communities, the physical structure of the rock is important. It determines if there will be crevices and ledges, how much soil can accumulate, and how much shelter there will be for plants and animals. Particularly distinctive are the bodies of granite and related rocks that formed deep underground and have no natural beds or joints to focus erosion. Instead, a process called exfoliation causes thin sheets to split off the surface and leave smooth, often rounded, rock outcrops with no crevices. This is the origin of the distinctive bare-faced domes such as Whiteside Mountain and Looking Glass Rock, as well as their lower-elevation counterparts such as Stone Mountain. With no crevices to provide deep rooting, only rootless and shallow-rooted plants can thrive, and the community is quite different from that on fractured rocks. We will talk more about the distinctive character of granite domes in the chapter on low-elevation rock outcrops.

As you might expect given their rugged nature, most high-elevation rock outcrops require some effort to reach. But the majority are on public conservation lands and can be visited. It is particularly important that you use care when you are in these communities, using trails and overlooks where they exist and otherwise staying on edges and on truly bare rock to avoid trampling plants.

The most spectacular collection of rocky summits in North Carolina is on Grandfather Mountain, where the swinging bridge affords easy access to some, and rugged trails with ladders lead you to more remote ones. Other rocky summits can be reached by trail at Mount Mitchell State Park, Mount Jefferson State Natural Area, Table Rock Mountain at Linville Gorge, and several Blue Ridge Parkway sites, including Waterrock Knob, Devil's Courthouse, and Craggy Gardens. High-elevation granite domes are concentrated in the southern mountains near the town of Highlands. An easy Forest Service trail provides access to the top of Whiteside Mountain, one of the largest. Other Nantahala National Forest domes accessible by trail in this area include Jones Knob, Whiterock Mountain, Scaly Mountain, and some in Panthertown Valley. Farther north, Looking Glass Rock is reachable by a Forest Service trail.

(opposite)
Heller's blazing star is a federally listed threatened species found only on rock outcrops at the tops of the highest mountains in North Carolina. Only eight populations remain.

Thin crusts of paintlike lichens, a cooperative growth of fungus and algae, are among the first living things to colonize the dry, sterile environment of bare rock.

Mountain Oak Forests

If you look up at the hazy green walls of North Carolina's mountains, or gaze down from some high vantage point across their wrinkled green ridges, much of what you see is mountain oak forest. While some of the communities we've already discussed are better known, the oak forests are the workhorses of our mountain scenery. They cover the majority of the natural landscape, draping the slopes and broader ridges at lower and mid elevations and climbing to the verge of the spruce-fir forests in places. Except at the highest elevations, any substantial hike in the mountains is likely to take you through oak forests. If you walk along a ridge trail, you may be in them for hours. If you hike a trail around a side slope, they may come and go, alternating with other communities as you cross the undulations of the topography.

As in most of our forests, once you are inside an oak forest, in all seasons but winter, you are in a private green world, closed off from the sky by a dense, leafy canopy. Green shadows and small sun flecks surround you, with larger patches of sunlight only in gaps where trees have fallen or died. You may find a dense thicket of dark green mountain laurel, a knee-high bed of huckleberries, a soft carpet of ferns, or a varied mix of herbs punctuated, in the right season, by the orange torches of flame azaleas, all just around the corner from each other. In the spring, the woods are alive with birdsong: the breathy squeak of the black-and-white warbler, the buzzing of the black-throated blue warbler, the bouncing flute-like song of the wood thrush, the drumming of grouse, and a tangle of others. Squirrels chatter from the trees. While you are less likely to see larger animals, a deer or even a black bear may be lurking around the next bend in the trail.

The oaks that dominate the canopy of these forests include a number of kinds, and you may see the broad, wavy-edged leaves of chestnut oak, the deeply lobed leaves of white oak, the many-pointed leaves of red or scarlet oak, or others forming most of the shade. Hickories, tulip poplar, maple, even pine may join them, though you won't see as many kinds of trees as are in cove forests. The canopy of those oak forests at the highest elevations may be composed of only red oak over many acres, while those at lower elevations may consist of nothing but chestnut oak.

Evergreen mountain laurel and, less commonly, rhododendron often grow beneath the canopy in mountain oak forests.

An old red oak is surrounded by younger
trees in this uneven-aged mountain oak
forest on Mount Sterling in the Great Smoky
Mountains National Park.

Mountain oak forests grow at intermediate moisture levels, between the moist cove forests and the dry conifer forests. Oak forests cover the most ordinary portions of the mountain landscape: the open slopes, the broader low ridges and peaks. Sometimes you can find them at higher elevations, too, but only on the warmest, west-facing or south-facing slopes.

If you hike much in mountain oak forests, you may notice some places that have dramatic variation in the size of the trees you see. Many trees in any given view may be about a foot or two in diameter; but many will also be smaller; and a few may be three feet in diameter or even larger. Mountain oak forests occur naturally as uneven-aged stands, where trees of many different ages grow intermixed. Though many of our canopy trees are capable of living 400 years or longer, some are always dying, from disease, lightning strikes, or storms. In any forest, one tree's disaster is another tree's opportunity. The only place young trees have any good chance of growing to adulthood is in sunny gaps in the forest canopy. So, continuous death of trees is accompanied by continuous birth of trees. You may have pictured an old-growth forest as one where all the trees are old. But a true old forest, like a human city, has old trees mixed with the young and middle aged, and is always full of death and birth.

This uneven-aged, old-growth structure is the natural state for spruce-fir forests, cove forests, and almost all of our forested natural communities, as well as oak forests. Why do most forests you walk in, even ones that seem mature, not look that way? Why do most forests have trees all about the same size, usually medium or small? Logging, which has occurred in most of our forests at some time, generally removes all the trees, or at least all the older trees, at once. Most of our forests don't have any large trees because there has not been time for trees to grow large after the last logging. Even most of our older forests have trees all the same size because they all got their start at the same time after logging. But in the mountains, you can sometimes find remnants of old-growth forest protected from logging by rugged terrain. These include small pockets as well as the famous larger examples such as Joyce Kilmer Memorial Forest. Many other mountain oak forests were only selectively logged. Some large old trees, too crooked or hollow to be worth cutting, remain to give you a hint of what old-growth forests look like.

Besides the oaks, one additional species of tree deserves special attention. Amid the smaller trees in the understory, you can often find saplings of American chestnut. They are recognizable by their oblong leaves with coarse, sharply sawtooth edges and a long point at the end. Chestnut was once a dominant tree in the canopy of most of these forests, so that the authors of older studies usually called these communities oak-chestnut forests. The large nuts they produced fed squirrels, bears, passenger pigeons, and Native Americans alike. What happened to them? Like the Fraser firs, the chestnuts were wiped out by a species that was accidentally brought from elsewhere, in this case a fungus called the chestnut blight. As the blight spread south through the Appalachians in the 1930s and 1940s, all chestnut trees died. However, the blight did not kill the roots of the trees. Some roots have lived on for the past seventy years, continuing to send up the sprouts that we see today in the understory of many mountain oak forests. The sprouts survive only a few years before the blight finds them, but they remain as a reminder of the magnificent trees they once were.

The loss of a dominant tree in forests over such a large area was a major ecological disaster. All of our mountain oak forest communities, even the most pristine ones, are heavily altered because of it. But there is also a lesson in the chestnut's demise about resilience in natural communities and about the value of diversity. The oak-chestnut forests had a number of tree species in them. Because other tree species were present that could fill the empty spaces left by the loss of the chestnuts, the majority of our mountain landscape remained forested, and most of the associated animals survived. This contrasts with the greater damage by the balsam woolly adelgid in the less-diverse spruce-fir forests, and even more so with the devastation of single-species pine plantations by southern pine beetles in other parts of the state.

Mountain oak forests are common throughout the mountain region of North Carolina, including the foothills, so examples are easy to find. The finest examples are those in large areas of old-growth forest, such as the unlogged portions of the Great Smoky Mountains National Park, Joyce Kilmer–Slickrock Wilderness (farther up the valley from the more famous Joyce Kilmer Memorial Forest), and Linville Gorge, all reached by trails. You can also hike for miles, if not days, mostly through oak forests, at South Mountains State Park, Gorges State Park, Hanging Rock State Park, Stone Mountain State Park, and countless places in the Pisgah and Nantahala National Forests.

Huge American chestnut trees once shared dominance with the oaks in mountain oak forests. Short-lived root sprouts are all you are likely to see of this species now, but rare individuals such as this one survive the blight long enough to become small trees and produce nuts. Efforts to breed blight-resistant American chestnuts may eventually allow this species to return to the forest.

Mountain Cove Forests

The cove forests may be the quintessential experience of the southern Appalachian Mountains. They best show the character that sets these mountains apart from both the northern Appalachians and from most of the mountains of the western United States. If you have hiked much in the mountains of North Carolina, you likely have found yourself in their world of green luxuriance, the damp exuberance of vegetation that stands out even among our many lush natural communities. Even in a typical second-growth cove forest, the trees tower, rising high to the dense canopy, their tops partly hidden by an understory of smaller trees. In an old-growth forest they are truly humbling. Bring your eyes downward from the trees and you will see similar intensity closer at hand, perhaps a shrubby wall papered with the big oval leaves of rhododendron or a waist-deep ground cover of flowery herbs. Despite all the places that animals can hide, you are likely to see something moving: a salamander twisting its way into the dead leaves, a brown millipede with yellow racing stripes patiently crawling along the trail, a snail gliding across a log, a bird flitting amid the trees, perhaps, with luck, a black bear. If you listen, you can hardly escape the sound of water. Creeks roar as they plunge down rocky ravines. Rain patters on the leaves and continues dripping hours after the clouds have moved on. And a musty smell of growth and decay joins the pervading dampness in the air.

Coves are small valleys, carved into the flanks of the mountains. You can find the natural communities we call cove forests in most valleys and ravines, as well as covering the lower slopes of ridges. Almost any time you visit, you can tell that these are damp places. They are moist not because of high rainfall and cool temperatures, as the higher peaks are, but because of topography. In fact, the rainfall in western North Carolina varies from about forty inches a year in some places to over eighty inches in others, and you can find cove forests that look the same in both. Water from the surrounding slopes funnels into coves, while the slopes shelter them from the full intensity of the sun and wind. You can see streams flowing down the valleys, but much water also stays below the surface, seeping in from above through the soil. Damp but not cold, wet but not saturated too long, coves are the most favorable places in the

(overleaf)
The new spring leaves of sugar maple and yellow buckeye trees begin to cast their shade on the lush wildflowers of this old-growth rich cove in the Pisgah National Forest. May-apple, larkspur, large-flowered bellwort, and violets are among the herbaceous plants visible in this scene.

mountains for plants to grow. And they are correspondingly hospitable to animals, fungi, and other living things.

Mountain cove forests are justly famous for their biological richness. The vegetation reflects the favorable conditions in the size of the trees, multiple layers of plants, and the number of species. Of the hundred or so kinds of trees in North Carolina's mountains, you can find most in cove forests: fluted columns of tulip poplar, dark-barked cherry, feathery-leaved ash and hickory, buckeye, maple, birch, oak, and scores of others. And no community offers such a variety of wildflowers spread through the seasons. Any time you visit, other than the dead of winter, you are likely to see a dozen kinds, from spectacular trilliums or lilies to the more subtle blooms of dolls'-eyes or sweet cicely. Amid them are dozens of more modest ferns, grasses, and other herbs. With the plants go the other living things in the community. Whether you focus on birds, salamanders, small mammals, insects, millipedes, mushrooms, or other odd fungi, whether you work to identify them or merely note their differences, you will find variety to satisfy you or overwhelm you.

While the moist conditions in coves are favorable for plants and animals, this does not mean that life is easy for those that live there. With the lushness of life comes much competition. Plants need light as well as moisture, and for all but the canopy trees, light is in short supply in cove forests. The umbrellalike shape of most of the herbaceous plants and many of the understory trees, with wide leaves spread in a single layer, is an adaptation for efficiently catching whatever flecks of light make it down to their level. Even most of the grasses in these forests have broad leaves rather than the narrow leaves typical in lawns. Though lush and exuberant, most of the herbs grow slowly. Young plants may take years to become large enough to flower. Many of the small plants on the forest floor send up their shoots each year from bulbs or roots that may live as long as the large trees. While cove forests contain more plant species than other mountain communities, many species are absent there or are more abundant in other communities because of the intense competition.

Another way that plants deal with intense competition is by being active at different times. Some herbs grow early in the spring and die back or hunker down without growing later in the season. Others grow slowly until summer or even fall then become large. If you visit a cove forest in the spring, you'll see violets, trillium, squirrel corn, may-apple, and many other kinds of flowers. Visit in midsummer and a different set of flowers and other herbaceous plants will predominate. Visit again in early fall and the herb layer will once again have been remodeled, with a new carpet of asters, goldenrods, and white snakeroot.

Mountain cove forests come in two distinctive forms, known to ecologists as acidic, or poor, and rich. Rich refers both to the diversity of plants and to the fertility of the soil. With

An early spring walk in a rich cove forest can reveal an abundance of beautiful species such as (clockwise from upper left) Dutchman's breeches, blue cohosh, sweet betsy, and large-flowered trillium.

Giant old-growth eastern hemlock trees tower over thickets of great rhododendron in this acidic cove in Joyce Kilmer Memorial Forest. The trees in this photograph have since been killed by the hemlock woolly adelgid.

48

moisture not a major concern, nutrient levels can have more influence than in most places. You can easily see the difference in the vegetation. Rich cove forests have lush herb layers with many species of herbaceous plants. Acidic cove forests are also lush but have dense thickets of evergreen shrubs such as rhododendron, or else a carpet of just a few species of herbs. Nutrient levels may be high either because the underlying rock contains more nutrients to be released into the soil or because of topography. Slopes that are concave, or bowl-like, in shape concentrate nutrients—more wash in than out. Slopes that are convex, such as those that occur in sharper ravines and on small spur ridges within coves, may still be moist but are less likely to be rich.

Old-growth tulip poplars in Joyce Kilmer Memorial Forest are among the largest trees in North Carolina.

While most of the mountain cove forests you visit are likely to show the exuberant lushness described here, you may encounter some that resemble ghost forests, with many of the trees dead. These communities are victims of the hemlock woolly adelgid, an introduced insect that began spreading through North Carolina earlier this century. It is proving as deadly to hemlock trees as the balsam woolly adelgid is to firs. Hemlocks are common in cove forests, especially in acidic ones. While cove forests have many other species of trees that will be able to fill in the space, the loss of the hemlocks will make a big difference. They are the only common evergreen tree in these communities, and the animals that seek winter shelter in their canopies will be left homeless. The insects that feed only on this species of tree, the decomposers that live on its distinctive needles on the forest floor, will share in its demise.

Mountain cove forests can be seen in most parts of the lower mountains; and most trails along creeks or to waterfalls will take you through some. Perhaps the most famous, justly so, is Joyce Kilmer Memorial Forest, where a visitor can walk trails through the massive trees of an old-growth forest. Great Smoky Mountains National Park has numerous trails through cove forests, including substantial areas of old growth in the backcountry. Some of the most extensive acidic cove forests, along with small rich pockets, are in the area along the South Carolina border, including Gorges State Park, Whitewater Falls and its gorge, and Ellicott Rock Wilderness. Linville Gorge also contains old-growth acidic cove forests, including extensive hemlock-dominated ones. Smaller but nice examples may be seen at High Shoal Falls in South Mountains State Park, Moses Cone Park, and a number of other places along the Blue Ridge Parkway.

Dry Conifer Woodlands of the Mountains

Compared to our hardwood forests, dry conifer woodlands are usually airy and well-lit places. Whether the trees are tall or short, the canopy tends to be open. Dark green pines stand stark against the sky, perhaps as tall, straight spires, perhaps as picturesque bonsai gnarls that evoke a life of lightning and wind. These communities often offer good views, with steep mountainsides dropping away on either edge to reveal the landscape beyond. Closer at hand, you may get a good look at a black-and-white warbler, normally hidden in the forest canopy, perched in the top of a tree below you. In a more distant view, you can often see the conifer woodlands as punctuation marks, lines of dark green amid the hardwood green or winter gray of a mountainside.

If you have visited a few of these communities, you know that open above doesn't necessarily mean open at your feet. Often a dense hedge of shrubs fences you onto the trail. If you want to go anywhere else, you must fight for your progress, wrestling stout branches, braving scratching twigs or the claws of a hidden greenbrier. But other woodlands are more passable, an inviting maze of grassy patches or rock beds amid the thickets. The shrubs can be a delight in their season as well—evergreen luxuriance in the gray of winter, mountain laurel walls covered with delicate pink flowers in spring, or blueberry bushes a glowing deep red in the fall.

Dry conifer woodlands are natural communities dominated by cone-bearing trees—usually pines but sometimes Carolina hemlock or eastern red cedar. As in other mountain natural communities, it is topography that creates the moisture conditions. Conifer woodlands are most often on sharp, narrow ridge tops, where rainwater quickly finds a way downhill and where the drying wind blows free. Some are on south-facing slopes, where the sun shines most strongly. Others occur where the soil is shallow and thirsty roots are blocked by bedrock close to the surface. These topographic controls on soil moisture trump rainfall. Dry conifer woodlands occur from the driest parts of the mountain region to the wettest, including in the rainiest places in eastern North America.

Besides dry soil conditions, fire is often an important ecological force in dry conifer wood-

Carolina hemlock clings to the steep, rocky cliffs of Bluff Mountain, a nature preserve managed by the Nature Conservancy.

lands, at least in the most common ones that are dominated by pines. Given these communities' occurrence on sharp ridge tops and upper slopes, the most lightning-prone parts of the landscape, this is hardly surprising. In addition, the tendency of fire to spread uphill makes it likely that the ridge tops will get burned by any fire starting in the neighborhood. Most pines are well adapted to fire. They have thick bark that lets them survive fires that would kill most

Stunted pitch pines cover the summit of Hawksbill Mountain in the Linville Gorge Wilderness.

hardwoods. Pitch pines, one of the common pine species of dry conifer woodlands, can even grow new needles from their bark if all their branches are killed by fire. At the same time, most pine seeds need bare mineral soil to germinate and are adept at colonizing newly burned areas. Table Mountain pine, another of the characteristic species of dry conifer woodlands, goes even further. Its seeds are held in sealed cones on the trees until the heat of a fire breaks the seal and releases them. Protected from the heat during the fire, they are released when

Dry Conifer Woodlands of the Mountains

the soil is bare and ideal for their germination. Other plants that grow with the pines are also well adapted to fire. Mountain laurel, rhododendron, blueberries, and the other shrubs readily resprout from their roots if their stems are killed by fire. The grasses, asters, sunflowers, legumes, and other herbs of dry conifer woodlands all grow more vigorously after being burned.

If you have hiked much in the mountain coniferous woodlands, you have probably seen places where most of the pine trees are dead. The cause is an insect called the southern pine beetle, which bores into the bark of pines and can kill them. Pine trees are normally good at defending themselves against southern pine beetles, and only trees weakened by lightning or some other damage are killed. But periodically, the beetles build up to very large populations that can overwhelm the defenses of healthy trees. This often happens during droughts, when all trees are a bit weakened. While most insects and fungi that kill large numbers of our trees are species introduced from elsewhere, threats native species are not equipped to resist, this is not the case for southern pine beetles. Given that they apparently have lived here for millennia, why are they so devastating? The answer is not entirely clear. In the past, our native pines have been able to produce new seedlings quickly enough to remain common. The current loss of pines is a concern, however, because they do not seem to be replacing themselves. Without the open space and bare soil opened up by fires, there is no room for pine seedlings after southern pine beetle outbreaks. Gradually increasing use of prescribed fire to replace the natural fires that are no longer allowed to burn may eventually reverse this trend.

Like the more common mountain laurel, Carolina rhododendron sometimes forms dense thickets beneath the pines.

Mountain conifer forests are widespread in the mountains, though they occupy a small minority of most landscapes. Excellent examples can be seen along the Blue Ridge Parkway north of Asheville, where they occur on the sharp ridges below many overlooks. Linville Gorge offers good examples, including extensive stands of Table Mountain pine on the west rim that were killed in wildfires in the last ten years and have regenerated. Some intact, old pine communities are also present within the gorge, such as around Babel Tower. Good examples can also be found along trails in Great Smoky Mountains National Park, Stone Mountain State Park, Hanging Rock State Park, and Gorges State Park. Carolina hemlock woodlands line the upper cliffs around Linville Falls and are present in Chimney Rock State Park and Hanging Rock State Park. Red cedar woodlands are rare and hard to reach. There are a few in Nantahala National Forest, and you can see one from a distance, perched at the top of the tallest cliff at Chimney Rock State Park.

Dry Conifer Woodlands of the Mountains

Mountain Bogs and Fens

In the landscape of the mountains, bogs and fens are unexpected natural communities. In a region known for its rugged and rocky terrain, they are flat and soft underfoot. Like the grass and heath balds, they are open to the sky, in a natural landscape that is almost completely forested. But, in contrast to the expansive balds, bogs and fens are small and secret spaces. You enter them through a dense thicket of shrubs, squeezing through the interlaced branches and the big evergreen leaves of rhododendron. Once inside, that green wall forms a backdrop to a hidden world. Most bogs and fens are complex, with herbaceous openings scattered in a matrix of shrubby forest or forming a labyrinth of connected openings. Many have patches of deciduous shrubs and scattered stunted trees within them rather than being a distinct meadow, shrubland, or forest.

Even if you have hiked extensively in the mountains, you may never have encountered one of these rare communities. Indeed, even if you know where they are, they are hard to visit. Many an accomplished biologist, used to finding a way in trackless forest, has gotten lost in a bog at some time. Beyond the shrub wall of an opening may be another meadow, or just an endless crawl through bushes. Try to follow a stream and you may find only little rivulets that appear out of nowhere, split and rejoin, and sometimes meander in every direction other than toward the nearest larger stream. If you walk into the heart of a bog, each step can be an adventure. Any one may land you up to your knees in soft muck. The plant you grab to catch yourself may be a greenbrier, a prickly swamp rose, or a poison sumac. But for the biologist, nicer surprises lurk as well. The next step may reveal an orchid flower or the trailing stems of cranberry, the big lush leaves of swamp pink or some other rare species seldom seen. If you're really lucky, it might bring a glimpse of the elusive bog turtle.

Bogs and fens are saturated wetlands with mucky soils. Because saturation leaves the soil without oxygen, which plant roots need, only specialized wetland plants are able to live there. Because mucky soils are low in nutrients, plants also must be able to survive extreme infertility. The result is a distinctive set of plants, many of which are not found in other communities. Little is known about the microbes and insects of the soil, but the distinctive environment

(opposite)
A beautiful mountain fen can be found on the Nature Conservancy's Bluff Mountain Preserve.

A thicket of red alder and swamp rose
surrounds this bog filled with skunk cabbage.

suggests they may be even more different from those of better-drained soils. These wetlands are usually found in the bottomlands of small or large creek valleys, but not right next to the creek. They are generally nestled against the base of the upland slope, where seeps feed water into them. The land often rises a little on the creek bank, partly trapping the water in the bog.

The most characteristic plant of bogs and fens is peat moss, or sphagnum. While species of moss are harder to identify than most plants, you can easily learn the distinctive look of sphagnum mosses. Their dense clumps of slender stalks with big shaggy heads, coarser than

Mountain Bogs and Fens

other mosses, are easy to spot. Also characteristic of bogs and fens are the grasslike sedges. They are some of the hardest plants to identify to species, but the variety of forms among their green seed spikes shows how many different kinds can be present. There are usually shrubs and small trees as well, but their numbers vary considerably from place to place. To an ecologist, the vegetation of bogs often appears to have a sense of humor. For example, pitch pine, the characteristic tree of mountain dry conifer woodlands, often shows up with its feet in the muck. Red spruce, too, mostly at home in the rocky soils of the highest peaks, makes a surprising appearance in many bogs. Next to these unexpected species may be plants more typical of coastal plain wetlands.

Mountain bogs and fens are one of the prime places to look for rare plants in the mountains. Like the high-elevation rock outcrops, they support a surprising number for their small size. They are rare habitats, with distinctive conditions that were more widespread during the ice ages. They are thus the last refuge for plant species that find the current landscape largely uninhabitable. Some of the rare plants are northern disjuncts, species that are more common far to the north, while others are endemic to the South.

Bogs are not known to have concentrations of rare animals, though this picture may change when the invertebrates are studied in more detail. One rare animal that is particularly characteristic, however, is the bog turtle. These small turtles bury themselves in the soft muck. They can be so hard to find that experts sometimes go years searching for them in a promising place before finding any. Unlike the plants, bog turtles do get around. They have been known to migrate up and down streams between nearby bogs, and occasionally to walk miles across uplands to get to new sites. But they are scarce, and getting more so, as commercial collectors carry them away and bogs are drained or destroyed.

The elusive bog turtle burrows in the soft, mucky soils of mountain bogs and fens.

To the scientist, bogs and fens pose a challenge beyond just finding your way around in them. They remain among the least understood of our natural communities. Even the question of whether they are bogs or fens is a matter of debate. You can usually find some sign of groundwater seeping into them, the defining feature of fens in the North. But, in most, the plants suggest they are highly acidic and have more in common with the rainwater-fed bogs of the North.

Mountain bogs and fens share with grass balds the paradox of being natural openings in a forested landscape that appear to be making every effort to grow up into shrub thickets or forests. They contain plants that need sunlight and could not survive in a shady forest, species that occur only in bogs and don't appear to move around readily. These plants may well have been in these spots since the last ice age. Yet invading woody plants threaten to make the bog plants' home uninhabitable in just a couple decades, and have already done so in some locations. The most characteristic animal of mountain bogs, the bog turtle, also needs sunlight and disappears when bogs become shrubby. Managers of bogs hold long debates over how to treat these communities and often put considerable effort into cutting the invading shrubs and trees to keep their bogs open. A few sites are not being invaded by woody plants, and these may hold the key to preserving these natural communities, but the answer has not yet been found.

The question is complicated by recent history. As you might expect of grassy openings in valley bottoms, bogs were used as pastures from the earliest days of European settlement. Even the periodic task of dragging out a cow that had become hopelessly stuck in the muck doesn't seem to have deterred early settlers. Most bogs and fens started filling with woody vegetation soon after grazing ended. So, were these places once kept open by the native elk and bison that roamed our mountains? Did the cows save the bogs after the elk disappeared? Should they be put back into bogs that are being invaded by shrubs and trees? Or did the cows destroy them, their heavy trampling and nutrient-rich droppings tearing up the distinctive soils and destabilizing a community that once could take care of itself?

Or maybe it was beavers. Before they were trapped out by the early settlers, their ponds must have been numerous in the valley bottoms. With beavers now returning, a number of bogs have ended up drowned by their ponds. But we also see bog vegetation hanging on at the edges of ponds, able to move back into the pond bed when the beavers move on. Bog turtles have found their way into other abandoned beaver ponds. Perhaps beaver ponds created bogs, or at least rejuvenated them periodically. But the increasing number of abandoned beaver ponds has not given us new bogs. So, the dilemmas and debates continue, with the fate of these rare communities possibly hanging in the balance.

It is not easy to see mountain bogs and fens. No public trails cross them, and the hazards of soft ground and getting lost make them unsuitable for casual visiting. Besides being hard to walk in, the soil is fragile and easily torn up if more than a few visitors pass through. Also, because many of the rare species in bogs are exploited by commercial collectors, owners justifiably try to keep their locations secret. Trail systems in the Pink Beds in Pisgah National Forest go by several small bogs and offer a good chance to see them from the edge. Most are several miles from trailheads. They are not marked but may be recognized as low-lying openings in the tree canopy with lush grassy-looking herbaceous vegetation. Small bogs also lie near trails in DuPont State Forest, along the southern part of the Little River. Several bogs are now in the state park system and may offer public facilities for seeing them in the future.

Cuthbert turtlehead in a bog in Panthertown Valley, Nantahala National Forest.

Upland Seepages and Spray Cliffs

To many people, the most memorable scenes in our mountains are the waterfalls. What an awe-inspiring drama to watch the smoothly flowing dark water shatter into white froth, leap through space along towering cliffs, and land with a roar in a pool of rocky foam. The clouds of spray that blow out from many falls heighten the feel of primal forces. You may not, at first, think of these wonders of water and stone as being natural communities. But look closely at the cliffs beside the plunging water and around the pool below, and you will see some of the rock covered with sodden and dripping mats of moss, liverworts, and algae. Little ledges and crevices will sport the tiny white flowers of saxifrage or alumroot, the shiny leaves of galax, or small ferns. Clumps of rhododendron, hemlock, or hardwood trees will emerge from the larger crevices. Some waterfalls have a grotto hidden behind them, and even under these dark overhangs the wet rocks are green with plant life. If you can get close to these places, you may see spider webs and daddy longlegs, slowly creeping pill bugs, salamanders, or the tracks of raccoons. Be careful, though, both to avoid falling on the slippery rocks and to avoid crushing the delicate plants with your hands. "Hand trampling" is a true threat and has led to the loss of rare species populations in some grottos.

In contrast to waterfalls, upland seepages seldom appear in tourist snapshots. Though they occur in all three regions of the state, these little pockets of wetness amid the upland forests most often go unnoted. In the mountains, where they are the most numerous, they may look like just a different kind of lush, herbaceous bed under the towering trees of a cove forest—a patch of delicate jewelweed, stinging wood nettle, or the dramatic umbrella-leaf. Higher, amid the spruce-fir forests or northern hardwood forests, beds of sedges may show their presence. In the piedmont and coastal plain, clumps of cinnamon ferns or royal ferns amid the blueberry bushes or brown leaves of the uplands may mark these communities. Look more closely and you will find soggy ground or trickling water, peat moss or other kinds of moss, and some biological surprises. Delicate orchids, pitcher plants, or sundews may nestle in the piedmont and coastal plain seeps, along with a diversity of sedges. Orchids, the showy flowers of turtle-heads or bee balm, the rare Gray's lily, insectivorous sundews, or even cranberries may inhabit the mountain ones. Others have simple vegetation, with just a few species of wetland ferns, sedges, moss, or liverworts. Hiding among them may be not only spiders and salamanders but a collection of distinctive seep-loving insects.

While the spectacular spray cliffs and the quiet seepages may seem an odd combination to share a chapter, their ecologies have many similarities. Both are wetlands in a landscape of drier communities. Both are wet because of slow, steady input of water, never deeply flooded but seldom or never dry. Indeed, the distinction between the two sometimes isn't all that sharp; if you look closely at many spray cliffs, you will see that seepage is part of what keeps them wet. But one thing that sets spray cliffs apart from most seeps is the high humidity in the air, in addition to the saturated ground. This makes spray cliffs particularly good places for mosses and liverworts. These tiny plants have no roots, and most kinds are very susceptible to drying.

Like the rock outcrops and bogs and fens, spray cliffs and seeps are home to a number of rare species. On the spray cliffs, the rare plants are mostly ferns, mosses, and liverworts. In counterpoint to the northern plants hanging on atop the rocky summits, some of these are shared with the wet tropical forests far to the south. With rock walls protecting them from the wind and with temperatures moderated by abundant water, these little pockets can be among the most stable of environments. These rare plants, and perhaps small animals as well, may have survived in these refuges since a time before the ice ages when our climate offered more extensive favorable habitat for them. Perhaps these same sites once offered refuge to other species that are now common, and thus allowed them to survive times of less favorable climate.

Upland seepages and spray cliffs are among the smallest of the naturally recurring assemblages of species and environment that we recognize as distinct natural communities. While a few cover an acre or more, most are small enough that branches from the trees in the adjacent forests shade them. In some ways, they also are among the most isolated of our natural communities. While a few border bogs or floodplain communities, most are far from other wetlands. They are surrounded by very different communities with which they share few species.

The small size and isolation of these communities raises an interesting ecological question. With a habitat so small, no species in a given seep or spray cliff can be very numerous, and many may have just a few individuals. One of the best-accepted understandings of modern ecological theory is the difficulty that small populations face. An inability to find mates that aren't close relatives, genetic drift, and random fluctuations in numbers all curse small populations. If another population is nearby enough that individuals, or genes in the form of

(opposite)
Ferns, mosses, and galax cling to the moist walls behind a waterfall in the Nantahala National Forest, while liverworts cover the ledge below.

pollen, can find their way back and forth, these problems can be overcome. But both theory and observation show that if a large population of some plant or animal species is reduced too much and becomes isolated, it is much more likely to die out, even if efforts are made to save it. And small remnants of once-large communities tend to lose species over time. Yet upland-seepage and spray-cliff communities, like the small and isolated high-elevation rock-outcrop communities, seem to be stable. There is every reason to believe these communities have been where we now find them for thousands of years. If they were losing species all that time, they should have run out by now.

The full answer to this paradox awaits the scientific studies of the future. Perhaps these communities aren't as isolated as we think. Perhaps wind-blown pollen and pollen-carrying insects do move from one seep to another. Maybe these communities aren't as stable as we think. Individual seeps tend to be quite different from each other. Much of that variation is clearly related to variation in wetness and other aspects of the environment; some may just reflect what species managed to find the small patch in the first place; but maybe some is the result of random loss of small species populations. Or perhaps these communities are composed of the species that have solved the problem of how to live in small populations. After thousands of years of being small, isolated communities, those species may be all that are left. There may be a big difference between populations that nature's evolution has adapted to being small and those that are abruptly made small by human actions.

Spray cliffs are easy to visit, or at least to see from a distance. Some of the most spectacular waterfalls have parking lots and overlooks, and many others have trails that lead to them. There are books describing the mountain waterfalls and how to find them. The largest and most distinctive spray cliff communities aren't necessarily at the biggest waterfalls; the shape of the falls and the cliffs matters as much as the height or amount of water. Perhaps the most spectacular spray cliff community is at Rainbow Falls on the Horsepasture River, in Nantahala National Forest adjacent to Gorges State Park. Another spectacular one, easy to see, is Dry Falls in Cullasaja Gorge in Nantahala National Forest. Upland seepages, though more common and present throughout the state, are harder to find. The largest collection of high-elevation seeps is in the Great Balsam Mountains, where several lie along trails in Shining Rock and Middle Prong Wilderness and in the adjacent area around Sam Knob. A large collection in Elk Knob State Natural Area may someday be accessible by trail. Several low-elevation seeps are visible from trails in Raven Rock State Park.

(opposite)
This upland-seepage community in Duke Forest is dominated by cinnamon fern. The surrounding trees are rooted outside of this small wetland, but their branches shade most of it. A small rivulet carries away the water that emerges from the ground.

Sedges, with their grasslike leaves, are common plants in upland seepages. This Howe's sedge, in Duke Forest, is one of many different species that you might find in these communities. Each individual seed is enclosed in a tiny sacklike membrane.

Piedmont and Coastal Plain Oak Forests

Just as in the mountain region, if you have spent time in the woods in central North Carolina, oak forests surely are familiar to you. As in most of our other forests, you may have noticed how a hike in one of these forests often brings a paradoxical feeling of expanse and enclosure. You are covered by a canopy that blocks out the sky, and sounds from outside are muffled. But most of these forests are not very brushy, so you can look off across the undulating land through the gray trunks. You feel like you are in a world apart, but one that is not necessarily that small or private.

In the piedmont, most of the landscape once supported oak forests. Their wide extent means they have borne the brunt of the bulldozing, plowing, and other activities that have left cities, fields, and successional pine forests the predominant scene in most of this region. But they remain the most abundant natural communities in the piedmont. In remaining natural areas, oak forests cover most of the broad upland flats and ridges, running down the drier slopes to interfinger with the moist forests in dissected lands near streams. The oak trees that define these communities tolerate the dry conditions that occur at times in these sites, and they do well in the periodic fires that once roamed the uplands. In the coastal plain, the story is different. Fires were too frequent even for the oaks, sandy soils too dry, wetlands more common. There, oak forests form little pockets amid dissected lands or lurk along with moist forests on isolated islands in swamps.

As in our other deciduous forests, the look of the oak forests changes through the year, offering new beauty with each season. It might be a pleasant, cool, shady green morning in summer, or a quiet world of soft grays and browns against the deep blue skies of a sunny winter afternoon. Autumn in these forests is remarkably prolonged, with understory trees turning deep red early in the season, hickories going brilliant gold in the middle, and oaks saving their pale or blazing reds for the last days before winter. Also visually distinct is the season of mid-spring. For a fleeting week or two, the partially expanded leaves and drooping flower catkins of the oaks paint the landscape in a mosaic of pastel colors before settling down into a uniform

fresh green. And the April oak forests are brightly decorated by the artists of the understory, the redbuds and dogwoods.

If you come from another part of the eastern or midwestern United States, the piedmont and coastal plain oak forests may look familiar to you. They are a part of the vast oak-hickory forest region that stretches westward all the way to Missouri. Their commonness may hide their complexity and diversity though. If you think an oak tree is just an oak tree, this is a great place to begin appreciating their variety. Here you can commonly find white, red, black, post, southern red, scarlet, and chestnut oaks. Though harder to identify, hickories are diverse as well, with five species fairly common. While the other layers of the forest are not generally very

Flowering dogwood and white oak are two of the most common trees of the North Carolina piedmont, but mature examples like these are rare. It takes many years for a dogwood to grow large in the shade of a forest. The flat-topped form of a wild, forest-grown dogwood is unlike the open-grown trees typically seen in gardens.

The mature oak forests at the North Carolina Botanical Garden in Chapel Hill are a rare example of land in the piedmont that has never been used for agriculture and has not had trees cut for many years.

dense or diverse, each has its own delightful species, from leaning sourwoods bending their way toward light flecks in the understory to the tiny, green and white, ground-hugging leaves of the rattlesnake plantain orchid. The characteristic animals of the oak forests are some of the most familiar animals in eastern North America: gray squirrels, white-tail deer, box turtles, black rat snakes, copperheads, the numerous warblers, chickadees, wood thrushes, vireos, and other familiar forest songbirds. These animals live in many kinds of natural communities, but the predominance of oak forests makes this their primary habitat. It is worth noting that other animals once were equally characteristic of the forests of the piedmont and coastal plain: black bears, mountain lions, wolves, bison. Black bears may still prowl the oak forests of the coastal plain, but the other large animals are long gone. Even many of the smaller characteristic animals, such as the elusive bobcat, thrive only in the largest forest remnants of the piedmont.

What became of these missing animals? People killed many of them, of course, out of hunger or mistrust, but there is another part of the story. Large animals, especially carnivores, must roam widely to seek food. Because each individual needs a lot of space, it takes a large

Downy arrow-wood, North Carolina Botanical Garden Nature Trail, Chapel Hill.

area to support enough individuals to have a viable population. Amid all the houses and roads, fields and shopping centers of the piedmont, the forests have not only been reduced in total acreage but have been broken up into many small pieces. We talked in the previous chapter about the woes of populations that are artificially reduced to small sizes and isolated from others of their kind. Many individual fragments are now too small to support such species. Thus, bobcats occur in some places but are no longer widespread. Black bears occur in parts of the coastal plain, with its large uninhabited wetlands, but not in most of the piedmont. Even without persecution, wolves and mountain lions would find few places in eastern and central North Carolina with large-enough expanses of forest to sustain them if they could return. The importance of large areas of forest is not limited to big carnivores, nor to oak forests. Many species tend to occur at low density on the landscape. They, too, are among the first to die out as forest patches get smaller and more fragmented. Thus, while any patch of natural forest is of value to many species, the bigger the patch and the better connected it is to other patches, the more species of the natural community are likely to thrive there.

Oak forests are common in the piedmont. Uwharrie National Forest has the largest expanses, and many trails provide easy access. Morrow Mountain and William B. Umstead State Parks have large areas, as do Duke Forest and Caswell Game Land. Most state parks in central North Carolina have at least some, as do many local conservation lands, such as Mecklenburg County's McDowell, Reedy Creek, and Latta Plantation Nature Preserves. In parks that are centered on rivers or reservoirs, such as Lake Norman, Eno River, Raven Rock, Falls Lake, and Jordan Lake, the oak forests are often mainly on the outer edges of the park. Oak forests are hard to find in the coastal plain, and, indeed, remaining examples are disappearing rapidly. Cliffs of the Neuse State Park and Croatan National Forest have some protected pockets that you can visit.

When new oak leaves and catkins emerge in the spring they briefly paint the landscape in a mosaic of pastel colors.

Moist Hardwood Forests of the Piedmont and Coastal Plain

If you have hiked in central North Carolina, there is a good chance you have experienced some of the moist forests. And you probably left with a pleasant impression of their charms. It may be those distinctive, pale gray, somehow elephantlike trunks of beech trees, the characteristic tree of these communities. It may be the beds of deep green Christmas ferns carpeting the ground amid the brown and gray of winter. You can relax in the particularly deep, cool shade on a hot summer day. Or thrill to finding the first showy wildflowers of early spring: the trout lily, bloodroot, and hepatica that herald the coming warmth from the midst of February. Pause to listen to the tapestry of birdsong in later spring and early summer, maybe picking out a favorite, like the lilt of the ovenbird or the four-note melody of the Carolina chickadee. You may be surprised to come across an unexpected thicket of mountain laurel, a bed of galax, even some Catawba rhododendron, plants more at home in the mountains than in the eastern lowlands. Or just enjoy the scene of a steep slope dropping away to a river below, or rising above a creek-side trail. If you manage to find one of the scarce examples in eastern North Carolina, you may be even more surprised to see such classically idyllic scenery amid the flatlands and swamps.

Ecologists call these natural communities mesic forests. "Mesic" is a word meaning medium or in-between, and to ecologists it is a technical term meaning not too wet, not too dry, but just right. As in the mountain cove forests, topography is the reason these places are moist. Sheltered ravines and bluffs that face north or east get less intense sun, so water lasts longer in the soil. Lower slopes catch water seeping in from above. In the warm climate outside of the mountains, mesic places are limited. Although common in the piedmont and sparsely present in most parts of the coastal plain, these communities usually occur as small pockets of a few acres at a time rather than in large expanses.

Mesic hardwood forests of the coastal plain are more variable than those in the piedmont. The coastal plain is largely flat, but here and there you can find bluffs and ravines along major rivers or creeks that support these communities. More surprising are "island" ridges deep

An unusually large American holly tree stands with beech in this hardwood forest on a moist flat in the coastal plain.

within swamps. These little pockets, often rising just a few feet out of the water, can be a particular delight to a sodden swamp slogger, or a swamp-dwelling bear. Other occurrences are on moist flats, where the water table is near the surface and a few inches of elevation make the difference between a wetland and a mesic community. Though these places may sometimes appear utterly flat, you can find the ridges by finding the beech trees. Only if you come when the water is just at the right level can you see that they are indeed on higher ground.

Besides offering water to plants, moist soils benefit natural communities in less direct ways.

Moist Hardwood Forests of the Piedmont and Coastal Plain

Along with being good for salamanders and snails, they favor fungi, bacteria, and many invertebrates. Active invertebrates and microbes cause dead leaves and wood to decompose quickly, releasing nutrients for use by living plants. Moisture also was an important influence on the forest fires that commonly played across the landscape not so long ago. These fires slowed down or stopped when they reached the moist, cool slopes, allowing the less fireproof plants and animals to survive there.

This pleasant environment supports plants and animals that cannot live in other, more stressful settings—species without special adaptations for drought stress or fire. On the other hand, without those environmental stresses and disturbances to limit growth, competition among plants becomes a crucial factor. The herbs and shrubs, even the trees when they are seedlings, must cope with the lack of light and with the dense tangle of roots taking up the moisture and nutrients from the soil. So, although it would seem that any kind of plant could live in such a place, only particular species are common. Living in mesic hardwood forests requires its own special adaptations. For instance, you can see that, as in the mountain cove forests, most plants have thin and wide leaves that are spread out with little overlap. This helps them make good use of the limited light.

Another adaptation is the tendency of many herbaceous plants to do their growing in the off season. In deciduous forests, there is more light on the ground in the short days of winter than in the leafy summer. There is also more moisture in the soil. Many plants have developed the ability to take advantage of this less competitive season, tolerating frost and freeze all winter with evergreen leaves, or growing in the warmer days but cold nights of early spring. At the extreme are the spring ephemerals, species such as trout lily, hepatica, bloodroot, and windflower. These little herbs produce many of the spring wildflowers that make these communities a delight in February and March. Their entire year's activity is confined to just a couple months, with leaves appearing quickly from underground bulbs and roots, followed soon by flowers. By the time the leaves on the trees are expanding in April, most ephemerals have produced their seeds, and the year's food supply is stored away underground. By summer, you can't find them at all. But they are not dead. Individual plants live through many years of these brief seasons in the sun.

One interesting aspect of many moist hardwood forests is the presence of disjunct plants and animals. "Disjunct" means they are small populations

Trout lily is a common spring ephemeral wildflower. It is one of the first flowers to bloom in spring and is gone without a trace by early summer.

(overleaf)
Beech trees dominate a moist hardwood forest on this north-facing piedmont slope. Beech often retain some of their brown leaves throughout the winter, providing a beautiful contrast with the evergreen Christmas ferns that so often occur beneath them.

The early spring flowers of bloodroot, waiting to open in the morning sunlight, can be found in moist forests where the soil is unusually rich.

that are separated from the rest of the range of the species by substantial distances. Just as the high mountains have species that are more common in the North, the cooler temperatures and abundant moisture in the mesic forests harbor species that are widespread in the mountain region but are otherwise absent in the piedmont and coastal plain. These include mountain laurel, which is found widely scattered well into the coastal plain. Catawba rhododendron occurs at several sites around Chapel Hill and Durham, but is otherwise mostly found only in the higher parts of the mountains. Other disjunct species include galax, trailing arbutus, and some of the lush herbs of rich cove forests, such as black cohosh and dolls'-eyes, along with red-backed salamanders and other small animals of the mountains. Some species, such as may-apple and bloodroot, make it to within a few miles of the coast. There are even some disjunct trees, such as the small population of Canada hemlock in Cary at Hemlock Bluffs State Natural Area, and the similarly unusual population of white pine in Chatham County at the White Pines Nature Preserve.

These populations are small, and they are many miles from the main range of the species. How did they get there? The likely answer lies in the same shifts of climate that we think moved the spruce-fir forests and other communities up and down the mountains and left alpine tundra species stranded on rock outcrops. During the ice ages, from 10,000 to several million years ago, the piedmont was as cold as the mountains are now. Many of the plants and animals we associate with the mountains were widespread in eastern and central North Carolina at that time. As the climate warmed, they were driven out by heat, drought, and fire. But in the small, cool pockets on north-facing bluffs, they had some shelter from these stresses. Why, then, don't you find them in all mesic hardwood forests? Between the last ice age and the present was a time when the climate was warmer and drier than it is now. Even in sheltered moist sites, only the luckiest populations were able to survive.

Moist forests are common in the piedmont. But remember these communities are usually small, and you may have to do some hiking and searching to find them. Look on the bluffs along larger creeks or rivers, especially on ones that face north. Most of the state parks in the piedmont region have some examples, including Eno River, William B. Umstead, Raven Rock, Lake Norman, and Morrow Mountain. Most other samples of the natural piedmont landscape have examples as well: Duke Forest, Mecklenburg County's McDowell, Reedy Creek, and Latta Plantation Nature Preserves, and many local parks. Moist forests are uncommon in the coastal plain, but a good place to see them is at Croatan National Forest's Island Creek Recreation Trail and at the Neuse River Recreation Area. A nice example can be reached by trails at Merchants Millpond State Park.

Moist Hardwood Forests of the Piedmont and Coastal Plain

Piedmont and Mountain Rivers and Floodplains

The piedmont and mountain regions are both blessed with many rivers and streams. Large, deep, muddy rivers slip quietly through wide valleys. Gentle creeks burble over gravel and cobbles. Frothing whitewater roars over boulders, crashing headlong down a gorge. A thin sheet of water slides silently down a slab of bedrock. A waterless, sandy channel bed waits for the next rain. An open pond created by a beaver dam stands where a stream once flowed, or a marsh stands where a beaver dam once held back a pond.

The floodplains of these streams vary in size from narrow strips to bottomlands a mile or more wide. They usually are lushly vegetated, with an abundance and high diversity of vines, shrubs, and herbs beneath a varied canopy. Along with the moist forests that so often adjoin them in the piedmont region, the floodplains often mount delightful displays of spring ephemeral wildflowers in February and March. You may find vast expanses white with spring beauties or covered with the mottled leaves and nodding yellow flowers of trout lilies taking advantage of the brief days of sun before the trees awaken. These floodplains may seem bursting with animal life as well, from lusty early-spring choruses of frogs in every puddle to the polyglot serenade of birds in the greening treetops. Find a muddy spot, and tracks may reveal the presence of more elusive creatures: raccoons, foxes, maybe a bobcat, mink, or otter.

Floodplains owe their character to the effect of flowing water, in both the long and short term. Floods in the piedmont and mountain regions tend to be quick and short-lived, and only a few low spots stay wet for very long. But those low spots, and the ridges and flats that separate them, were sculpted by flowing water. The soil itself was washed in from elsewhere. You can see evidence of this process in the form of fresh layers of mud on the ground after floods. Piles of dead leaves and sticks on the upstream side of trees or other obstacles attest to the ability of flowing water to rearrange things. The young deposits and continual input of new material make for very fertile soils.

If the water is clear, you may be able to see some evidence of the aquatic community that inhabits the stream itself. Sit on the bank and you may see crayfish or insects crawling amid the

cobbles. Small fish are often visible, and sometimes larger ones such as the trout of clean, cold mountain streams. Scoop a net or turn over rocks of a clear stream and the larvae of many kinds of mayflies, stoneflies, or caddisflies may be revealed. Harder to find, but an important part of the fauna, are mussels, quietly filtering their sustenance from the water they pump through their shells. Often, the only sign of them you'll find is piles of shells on the bank, left by the muskrats and raccoons that eat the mussels.

You will notice that no plants have been mentioned in talking about the aquatic community, and, indeed, you may see none. If you were to look through a microscope, though, you would see the green filaments of algae or the exquisite, patterned shells of diatoms. In contrast to all of our terrestrial natural communities, in aquatic communities the plants tend to be microscopic, short-lived algae, and the living things that form the structure of the community are animals. Indeed, though there are some natural beds of aquatic plants in streams, usually when you start seeing large amounts of algae it is a sign that the community is degraded by pollution or is badly out of balance.

Streams are abundant in both the piedmont and the mountains. If you try to go in a straight line in any direction, you seldom get far without encountering one. Look at these regions from an airplane, or on a detailed map, and you see them forking, and forking again, their branches reaching up into every part of the landscape except the highest ridges. The abundance of streams shows the great geologic age of these upland areas. This contrasts with the young surface of the eastern coastal plain, where streams are far apart and have few branches.

Though piedmont and mountain streams and rivers are abundant, finding their natural communities in good condition is difficult. This is especially true of the larger rivers. In the mountains, the valleys of large rivers are the only flat land. Though floods can be swift and disastrous, river bottoms are often the only place for fields, roads, even towns. Alternative building sites are more common in the piedmont, but the fertile floodplains of large rivers are prime places for farms. Such uses go back farther than you may think. River valleys were the principal locations for the fields and major towns of the Native Americans long before Europeans arrived on the scene. Dams have taken a major toll as well. While large reservoirs such as those on the Catawba and Yadkin River are the most visible and perhaps the most lasting, a walk along the Eno River, Crabtree Creek, and countless other piedmont streams and rivers will reveal the remains of old mill dams and their reservoirs. Though these ponds have long been drained, they have had lasting effects on the streams and floodplains.

Even where floodplain communities have escaped the plow and bulldozer, their green lushness often hides an additional ecological problem. Some of the most abundant plants don't really belong there. They are species that were brought by modern people from far away.

(opposite)
Morgan Creek is lined with river birch as it meanders by Mason Farm Biological Reserve in the piedmont near Chapel Hill.

While most such introduced species don't "escape" beyond gardens or spread much beyond disturbed areas, a few are invasive. Without the diseases, specialized herbivorous insects, and competitors that keep them in balance in their homeland, they run rampant. Then they can crowd many other species out of natural communities. While such invasive "exotic" species are even more of a problem in many places in the world, the place they cause the most harm in North Carolina is in piedmont and mountain floodplains. It is hard to find a floodplain where you don't see them. Dense tangles of Japanese honeysuckle vines choke young trees. Thickets of the evergreen Chinese privet or silvery-leaved autumn olive dominate the shrub layer instead of the native spicebush. Beds of Japanese stilt-grass drape their thatch over the remains of a once-diverse herb layer. Dense beds of common chickweed squeeze out the spring beauties in March. It isn't just the floodplain plants that are victims. Piles of the shells of Asiatic clams show that they often vastly outnumber the native mussels in the water. It probably is not an accident that invasive plants are more numerous in floodplains than in other kinds of natural communities. The soils are among the most fertile; floods help spread seeds; and the periodic scouring and movement of soil by running water creates open spaces for them to get a foothold.

Invasive plants are a dilemma for natural-area managers. You can usually kill these plants with herbicide or digging. But it is hard not to kill the native plants at the same time, destroying the natural community you were trying to save. Sometimes a biocontrol can be found—a specialized herbivore or disease that kills the invasive species while sparing the natives. But history is full of examples of biocontrol attempts that have introduced a worse problem than the one they were trying to fix. In general, only skilled weeding or very careful application of herbicide to individual plants truly works. Sometimes only careful intensive management can return a floodplain community to a natural state once it has been invaded.

The aquatic communities of rivers also are susceptible to ecological damage in other indirect ways. Sewage is usually well treated these days before being dumped into rivers, but even treated sewage carries nutrients at unnatural levels. And runoff from uplands also brings the river the accumulated mud, fertilizer, pesticides, oil, and miscellaneous spilled chemicals of construction sites, lawns, golf courses, and farms. Modern urban life has brought more variety to these insults, but erosion of fields in the 1700s and 1800s had already buried many streams in mud. More subtle, but also crucial, clearing or building anywhere in the watershed causes more rainwater to drain rapidly into streams instead of soaking into the ground. Having as little as 10 percent of the watershed's ground covered with pavement or buildings is enough to badly degrade a stream, making the channel unstable and increasing the load of mud.

(opposite)
Mountain streams and rivers such as Looking Glass Creek in Pisgah National Forest rarely have much floodplain because of the steep terrain.

For all these reasons, the aquatic communities of our rivers, especially those in the piedmont rivers, are some of the most degraded and threatened in the state. Many species of aquatic insects are so sensitive to pollution that the state's Division of Water Quality uses them as a measure of how clean the water is. Even more sensitive are the mussels, who filter their food out of large amounts of water. No group of animals or plants has as many of its species endangered in North Carolina as the freshwater mussels.

Floodplain natural communities play a special ecological role in the landscape. Their lush vegetation slows floodwaters and filters sediment and pollutants out of them, helping prevent natural and human disasters downstream. And in heavily settled parts of the piedmont, floodplain forests are sometimes the last natural vegetation. Even as their lush greenness and natural hush may provide people their only respite from the city, they are often the only habitat left to shelter deer, foxes, songbirds, and other characteristic wildlife. Because floodplain communities tend to form continuous chains of forest along miles of stream, they also can form landscape connections that let animals move between patches of remaining forest. Indeed, they are often the routes left for animal migration in fragmented landscapes. They make it possible for some species to survive in places where no single patch of forest is large enough to support a viable population.

Floodplain communities and aquatic communities in varying condition are easy to visit in most parts of the piedmont. Most of the state parks have good examples of larger rivers or medium-size streams, including Eno River, Raven Rock, William B. Umstead, Morrow Mountain, South Mountains, Hanging Rock, Haw River, and Deep River. Many larger county and city parks also contain streams, as do most parts of Uwharrie National Forest, most of the large game lands in the region, Duke Forest, and a number of private preserves. Most of these floodplain communities have invasive plants but have relatively mature forest. Intact floodplain forests are scarce in the mountains, where most small and medium streams don't have floodplains and where larger floodplains are heavily altered. The best remaining example is along the Little Tennessee River, in the Needmore Game Land. Another good place to see floodplain communities is on the French Broad River in Pisgah National Forest, downstream from the town of Hot Springs.

Spring beauty, one of the first wildflowers to appear in the early spring, sometimes carpets the ground in piedmont floodplains.

Piedmont and Mountain Rivers and Floodplains

Low-Elevation Cliffs and Rock Outcrops

Low-elevation cliff and rock-outcrop communities are a diverse group of natural communities that share open vegetation and substantial amounts of bare rock. In the naturally forested world of the piedmont and mountains, they often stand out as the most memorable of communities. Some, the lower counterparts of the high-elevation rock outcrops, offer expansive, sun-drenched views from peaks or upper slopes. Others are valley-side cliffs that brood over rivers or coves and provide shady retreats. And others are flat floors that hide in secret chambers amid the forest. Like the high-elevation rock outcrops, many are inviting and promise good views. And as with high-elevation rock outcrops, there is reward to you, and to the natural community, if you watch your step and enjoy what is underfoot as well as the distant vista.

These natural communities are quite variable, but one common feature is irregular vegetation structure. Trees, shrubs, herbaceous plants, mosses, and lichens, all grow where they can, where a mat of soil or a crevice allows an individual or a small patch to take root. There is not the distinct canopy or herbaceous layer that you find in most of our natural communities. If you look closely at the species of plants, you will find that they are an interesting mix. There are usually some tree seedlings, and often some galax, Solomon's seal, or other herbaceous species from the nearby forest. They may not really be at home but are common just because plenty of seeds land here. Weedy plants such as ragweed and broomsedge, loving light and adept at dispersing their seeds to clearings far and wide, often find their way to the rocks. Species such as little bluestem, coreopsis, and eastern red cedar favor many different kinds of natural sunny places. And there may be specialist species, which need the peculiar conditions of particular kinds of rock outcrops and grow nowhere else. Given the lack of cover, there may be few kinds of animals that spend most of their time on the rocks. But the abundant sunlight makes them attractive places for basking snakes and lizards. And, if you look carefully, you can find salamanders, spiders, even bats, lurking in the moist crevices of more sheltered rocks. There are a few specialists among the smaller animals, too, such as grasshoppers perfectly colored to hide on lichen-covered rocks.

(overleaf)

One of the most spectacular low-elevation granite domes is Stone Mountain in Stone Mountain State Park.

Given the geologic history of the southeastern United States, it is somewhat surprising that there are any rock outcrops at all. Though the piedmont and mountains are mostly composed of hard rocks, many millions of years of abundant rain and warm temperatures have taken their toll on them. While you might think that all our rain and running water would have washed away the soil and left much bare rock behind, a process called chemical weathering has happened even faster. The moisture and natural acids seeping through the soils have converted the most common minerals, the feldspars and micas, to clay. Dig beneath the soil and you will find several feet of saprolite—"rotten rock" you can cut with a shovel—before you hit hard bedrock. So, rock outcrops occur only where rocks are unusually resistant to chemical weathering or where unusual erosion has bared them. Quartzite and hard rhyolite are particularly good at forming rocky summits, while undercutting by streams or rivers can form cliffs that reveal various kinds of rock.

One type of rock that is particularly distinctive is granite and its close relatives. We discussed the process of exfoliation in the chapter on high-elevation rock outcrops. Exfoliation acts similarly at lower elevations, forming bare-faced domes in the lower mountains and foothills. In the flatter piedmont, exfoliation produces level, pavementlike expanses called flatrocks. Though they are rare, these granitic outcrops have been of particular interest to scientists, and numerous studies have been done on them since the 1940s. While most other rock outcrops have crevices in which plants from herbs to trees can potentially take root, granitic flatrocks and domes have almost none. Instead, soil accumulates slowly in thin mats. Moss growing on the bare rock traps washing sand grains and blowing dust. Eventually enough accumulates that small annual herbs can grow. They trap more sand and dust and add their own dead leaves and roots. When the mat is thick enough, larger perennial herbs and grasses can grow. When more soil has accumulated, shrubs and even trees may be able to survive. You can often see the results of this process as a set of rings around larger mats, thick enough for a tree or clump of bushes in the middle, thinning to a grass ring near the edge, and finally to an annual herb ring or just moss. You may also see small mats that are just starting, which may have only the thinner zones. Different kinds of mats may form on convex rock, where water runs off, than in small depressions where water accumulates. And the edge of the outcrop, where the adjacent soil ends and its moisture seeps out, often has a distinctly different, wet kind of mat, with peat moss, sedges, and other wetland plants. The flowery and mossy mats give these communities a spectacular beauty like no other.

Because of these distinctive processes, granite flatrocks and domes have some of the most distinctive and rare plants among the low-elevation rock-outcrop communities. You'll see

(opposite)

Fragile, bright red, succulent mats of elf-orpine cover parts of the granite flatrock at Mitchell Mill State Natural Area.

Low-Elevation Cliffs and Rock Outcrops

The unique, wave-washed limestone bluff on the shore of Lake Waccamaw is lined with Venus'-hair fern, a rare species in North Carolina.

the bright red, succulent mats of elf-orpine and beds of the white flowers of single-flowered sandwort only on granite flatrocks, for these species grow nowhere else. Likewise, the granite dome goldenrod is a species found only on granite domes.

If this kind of natural succession and soil accumulation continued unabated, all the granite outcrops would long ago have been covered. But the soil mats of granite domes and flatrocks

Low-Elevation Cliffs and Rock Outcrops

have a limited life span. If a windstorm blows down a tree in an old mat, its falling pulls up the mat and leaves bare rock behind. In droughts, the soil mats dry out and any trees growing in them die. The dead trees eventually fall, once again pulling up the mat and destroying it. On the steeper granite domes, mats eventually fall off under their own weight. And, in the end, exfoliation will cause the rock itself to shed its surface and renew the outcrop.

North Carolina's rock outcrops are, almost by definition, confined to the piedmont and mountains. The coastal plain has virtually no rock. But the rare exceptions are interesting natural communities in their own right. A few river bluffs have exposed vertical faces of the partially consolidated clays and sands that lie beneath the coastal plain. Though the rock is not very hard, these are similar to piedmont cliffs in most ways. More interesting are the few outcrops of limestone in the outer coastal plain. Though typically too small to form open cliffs, the unusual chemistry of the limestone supports some plants not found elsewhere in North Carolina. A unique wave-washed limestone bluff on the shore of Lake Waccamaw, for example, has Venus'-hair fern, a plant more at home in the canyons of Arizona and Utah. The shaded limestone outcrops of Island Creek, near Pollocksville, support two other species of rare ferns, along with columbines and beds of unusual mosses.

Though rare, many of the best examples of low-elevation cliffs and rock outcrops are protected and easy to visit. Hanging Rock, Pilot Mountain, and Crowders Mountain State Parks owe their fame to rugged quartzite rocky summits. Raven Rock State Park's signature feature is a tall, dark, overhanging cliff along the Cape Fear River, and Cliffs of the Neuse State Park harbors the state's best example of a coastal plain cliff. Some of the most extensive low-elevation cliffs line the walls of Linville Gorge. You can climb over some of the most spectacular low-elevation granite domes in Stone Mountain State Park, while an even larger collection is scattered around DuPont State Forest. You can walk to the best coastal plain limestone outcrop on the Island Creek Recreation Trail in Croatan National Forest. Granite flatrocks are harder to visit. One of the best publicly accessible examples is Forty Acre Rock, over the line in South Carolina. One of the best examples in North Carolina lies in Mitchell Mill State Natural Area, and several more are protected in Wake County parks and private preserves. These are not designed for ready public access, because the lichen, moss, and herb mats are fragile and could easily be destroyed by too many people walking on them.

Elf-orpine grows with reindeer lichen on a granite flatrock at Mitchell Mill State Natural Area.

Piedmont and Mountain Glades and Barrens

Glades and barrens are in-between natural communities. Like the cliffs and rock outcrops, they are dry, naturally open places amid the forests. Glades and barrens don't often offer expansive views, but they have open, often patchy vegetation that makes them attractive places to look around. Answer their invitation to explore, and you may wander through the curly leaves of oatgrass, shaggy little bluestem, or the shimmering sea of springtime needlegrass. In the mountains, delicate beauties such as shooting star may nestle amid the grass beneath red cedars or chestnut oaks. In the piedmont, big yellow sunflowers, yellow prairie dock, the incomparable purple heads of coneflower, or the purple spikes of blazing star may stand like fellow wanderers, rising above the summer grass beneath a gnarled canopy of post oaks. Try to imagine the bison and elk that must have once favored these grassy places amid the surrounding forests, or the long-gone gray wolves and mountain lions that surely stalked them.

Glades and barrens naturally have open vegetation, but unlike rock outcrops, plant cover is substantial. Many of these communities have irregular vegetation structure—clumps of trees alternating with shrub thickets and grassy beds. But others are more uniform, with an open tree canopy covering a grassy groundcover over a broad area. They are open because some aspect of the site, something other than wetness or complete lack of soil, prevents trees from growing there as densely as in a forest. For some, the cause is rock lying beneath a thin layer of soil. Trees may be able to grow in the deeper pockets but any that take hold in other parts will find too little reserve of moisture to survive the droughts that will inevitably occur. The other typical creator of these open communities, especially in the piedmont region, is a clay hardpan. Hardpans form when the combination of rock type and topography are right. Unlike the typical southern red clay, it forms a subsoil so dense that roots can't readily penetrate it. Though water often sits on the surface after rains because it cannot soak in, these soils become very dry during other times. Only the most drought-tolerant trees, such as post oak, can grow in them. Add the occasional fire, as once was common, and these trees have a difficult time forming a closed canopy.

Management by the
North Carolina Botanical
Garden at Penny's Bend,
including prescribed burning
and clearing of invading
trees, has allowed this
population of eastern prairie
blue wild indigo to thrive.

89

(opposite)
The North Carolina Plant Conservation Program has opened the forest and improved the habitat for the federally endangered smooth coneflower at the Eno Diabase Sill Plant Conservation Preserve.

Though glades and barrens are stressful places for trees, what is bad for the trees is good for the smaller plants. The lack of shade and the reduced competition from trees makes room for a host of herbaceous plants that can tolerate the dry conditions but cannot thrive in the shade of a forest. At their best, these communities are rich in plant species, including many rare ones. A number of rare butterflies and moths depend on the plants of these communities. Some of the birds we now associate with fields and clearings may once have had glades and barrens as their primary home.

Finding these communities at their best can be an impossible challenge. While glades that are created by shallow soils over rock may have changed little over the years, the same is not generally true for the more extensive barrens of piedmont clay hardpans. Here, the fires that once had free play over most of North Carolina were particularly important. As they burned through most of the landscape, these low-intensity fires did little harm to the forests. But the same fire could have much greater effect in the dry soils and denser grass of a glade or barren. In a place already stressful to trees, a little fire would have created a canopy even more open than what we now see. And a more open canopy would have allowed even denser grass to flourish. Just how much more open the canopy was is somewhat uncertain. Perhaps something as open as our longleaf pine savannas. Perhaps the most frequently burned and driest were largely treeless. Early explorers in the piedmont, while traveling through a landscape that was substantially forested, wrote of areas open enough that they called them prairies. Since these explorers usually followed the trails of Native Americans and went from one of their villages to the next, many of these open areas were clearly abandoned fields. But others may well have been permanent natural grasslands or savannas, maintained by fire and hardpans.

Another clue could come from the plants themselves. The piedmont region has a number of rare plant species that need abundant sunlight. Many of these plants are closely related to species of the midwestern prairies. Most now hang on along mowed roadsides, power-line corridors, and old pastures—places with abundant light but without too much soil disturbance. Many, such as the prairie dock, smooth coneflower, and blue wild indigo, occur only in areas with clay hardpans. Some, such as Schweinitz's sunflower and Georgia aster, occur near hardpans and also in regions with other evidence that fire was once particularly frequent. Growing with them are more widespread species shared with the Midwest, such as big bluestem and Indian grass. These collections of plants hint that a diverse grassy com-

Schweinitz's sunflower is one of a suite of rare plant species that hang on largely in artificially open areas because piedmont barrens have become too shady.

Piedmont and Mountain Glades and Barrens

Sparse pitch pine and prairie grasses dominate this barren at Buck Creek in Nantahala National Forest. Serpentine, a rare type of rock, creates the environment for this unique natural community.

munity occurred there not so long ago. While these remnants may, at first glance, look similar to the short-lived successional weeds and grasses of abandoned farmland, they represent something quite different. The plants of the piedmont barrens are mostly long-lived species that do not readily seed into disturbed areas but would form a stable community once established. While most of the piedmont barrens have been lost to development and other land conversion, almost all of the remainder have been lost beneath increasingly dense tree canopies. In a few promising remnants, land managers are using prescribed fire and tree thinning in an attempt to restore these communities.

Visiting piedmont and mountain glades and barrens is difficult. Though a number are on public lands, none are marked and few are accessible by trail. Black Mountain/Parker Knob and the Pinnacle in the Roy Taylor Forest area of Nantahala National Forest are a couple of mountain glades that lie near trails. The largest collection of rocky glades in the mountains may be in the new Chimney Rock State Park, and public trails may eventually allow access to some. A large glade is visible from I-26 on the face of Tryon Peak in Polk County but is not readily reachable. The largest collection of piedmont rocky glades may be in Uwharrie National Forest, but none are currently reached by trails.

Seeing the remnants of hardpan barrens in the piedmont is almost as difficult. One of the best places is Penny's Bend, near Durham. Short public trails take you through old pasture areas where a number of the rare plants are present, and prescribed fire in nearby areas is restoring some of the natural vegetation structure. Several other remnants are protected by the state Plant Conservation Program but may be visited only with permission. There are other places around the piedmont that, while not true restorations of glades or barrens, have plantings of prairie grasses and other characteristic plants that occurred in them. These can give you some feeling for what these natural communities once looked like.

Eastern bluebirds occur mainly in artificially cleared areas today because open natural communities have become too dense.

Piedmont Upland Swamps and Pools

Like rock outcrops, glades, seeps, and mountain bogs, piedmont upland pools and swamps exist as small patches tied to special environments. They occur as pleasant little surprises amid the expanses of forests that make up the piedmont's remnant natural areas. You may find them most easily in the late winter or early spring, when a lively chorus of frogs calls to you. You can follow their song until you come upon an unexpected shine of water amid the brown leaves, with the gray trunks of willow oaks or the pale bark of overcup oaks rising from their reflections in the pool. If you look more closely, picking carefully among the strands of greenbrier that often line the edges of these communities, you may see clumps of moss at the water's edge, basking brilliant green in the sunlight. Like the ephemeral herbs of the moist forests, they are taking advantage of this season in the sun before the shade of summer returns. If you look more closely still, you may find masses of translucent eggs: the up-and-coming next generation of frogs or salamanders. Look very closely, and you may find the waterlogged leaves alive with tiny bugs, or even the clearly recognizable forms of fingernail clams. If you return in later summer, the water will likely be gone, the shallow depression almost invisible. The tadpoles will be grown and departed, and the clams and bugs buried in the earth. Only the presence of the willow oaks, greenbriers, sedges, and a few other wetland plants will tell you that this is an extraordinary natural community.

Upland swamps are a rare paradox of a community. Wetter, open upland pools are even rarer. It is unclear why there should be a basin or depression that would hold water amid the rolling, dissected lands of the piedmont at all, and the cause of such landforms is not well studied. To heighten the paradox, they generally occur not with the moist forests, but with the driest of upland communities. Most are on flat landscapes with clay hardpans, where they sometimes border the barrens. The willow oaks that most typically dominate the swamps often stand shoulder to shoulder with post oaks, among the most drought-tolerant of our trees. Other upland swamps and pools lie in folds of ridge tops, amid rocky dry oak forests. In general, they have little watershed and no input streams. The water that fills them is rainwater.

(opposite)
A still pool of water reflecting the moss-covered buttresses of overcup oak is an unexpected sight on top of a hill in the Uwharrie National Forest.

94

The water is generally blocked from draining into the ground by a hardpan or bedrock, and it stands in the basin until evaporation and plant roots dry it.

These pools are particularly important from the viewpoint of the amphibians. Most frogs and salamanders, though they spend their adult lives in various upland and wetland communities, need standing water for their eggs. They seek out pools where fish are unlikely to prey upon their tadpoles. These small bodies of water, remote from streams and periodically dry, are ideal. Along with the occasional fish-free piedmont and mountain floodplain pool, they are the most important breeding habitat for many species, from the common upland toads and frogs to the rare four-toed salamander.

As with all communities that occur as small patches, the surrounding landscape is very important to piedmont upland swamp and pool communities. This may seem a bit surprising, since they are so different from the surrounding uplands that virtually no plants are shared. In addition, often little water runs off from the surrounding uplands into the basins. But the importance is dramatically illustrated on rainy late-winter nights when salamanders converge on these communities from all directions. The breeding amphibians that are such an important part of these communities can be present only if the upland habitats where they spend their adult lives are left habitable and if they can get from one habitat to the other. Loss of the upland woods, or an impassable barrier such as a developed area or major road, can leave a pool bereft of most of its amphibians.

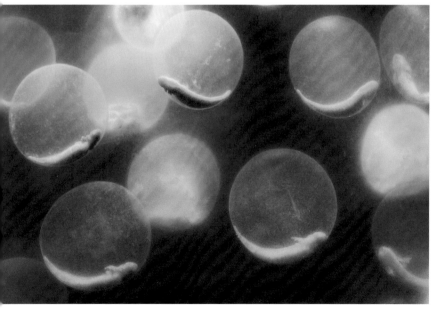

Temporary upland pools are essential to the lifecycle of animals such as the spotted salamander that must lay their eggs in water free of the predators found in more permanent ponds and streams.

Piedmont upland swamps and pools are not easy to visit casually. Though some examples are protected on state and federal conservation lands, none are marked for visitors and few lie along trails. You must either know where to look or be lucky. There are a number in Uwharrie National Forest, and you may come across one if you wander widely. One of the easiest to find is just south of Pleasant Grove Church in the forest unit south of Troy. Another is east of the Arrowhead Campground, north of its access road. One of the state's best upland swamps occurs on the Caswell Game Land, north of the appropriately named crossroads of Frogsboro. Some of the largest piedmont swamps are in Charlotte, including several owned by the Catawba Land Conservancy but not generally open to the public. Others are present in Mecklenburg County's McDowell and Flat Branch Nature Preserves. Duke Forest's Gate 9 Pond, near Durham, is an example that has been the subject of a number of research projects.

Piedmont Upland Swamps and Pools

Dry Longleaf Pine Woodlands

If you are a North Carolina trivia expert, you may know of the longleaf pine as the state tree. If you are a history buff, you may know of it as the basis of the state's largest industry in colonial times: the collection of its sap for pitch and turpentine, known as naval stores. Or you may simply remember it as one of the most attractive trees we have, with its clear straight trunks, pomponlike clusters of amazingly long needles, and giant cones. The communities it presides over are also inviting places. The typical wide spacing of trees—an open woodland rather than a dense forest—gives these communities an airy, sunlit character. Even the shade beneath each tree is light, as sun gently sifts down through the open crowns. The individual tufts of wiregrass beneath your feet blend into a lawn stretching into the distance beneath the trees. Look closely and you will find many other small plants amid the grass clumps: asters, sunflowers, many versions of pea flowers, ankle-high huckleberries. Knee-high scrub oaks may also lurk amid the grass, or larger ones may form an understory. The slightest breeze sighs with disarming gentleness through the pines, punctuated by the squeaky call of a nuthatch or maybe the distant "bob white?" of a quail. With luck, you might see a fox squirrel, unexpectedly large, bounding through the grass. With a little care to avoid the occasional poison oak and the stinging hairs of tread-softly plants, you can sit and feel yourself relax amid the peaceful waving grass and shimmering pine needles of a sunny day.

Dry longleaf pine communities can inhabit virtually every kind of upland site in the coastal plain. Indeed, they once covered almost all of the well-drained lands throughout that region. Like a wave rolling in from the ocean, they washed up into the edge of the piedmont region as well. While most of our plants can live in only a narrow range of moisture conditions and nutrient levels, longleaf pine trees are nearly indifferent to such concerns. Their remarkable range of tolerance makes them at home in places that are sodden for much of the year, and also in the driest places in the state. They also range from rich, fertile, loamy soils to the most sterile of sands. The most loyal companion of longleaf pine, wiregrass, also thrives in this broad range of conditions. With such abilities, what could go wrong in life for these species? Their major weakness is that they cannot tolerate competition. They must have the sunlight and

Longleaf pine and wiregrass dominate the scene in much of the Sandhills Game Land. The presence of tall flowering stalks on the wiregrass indicates that this area burned within the last year or two.

open space that characterizes these communities. A cover of other kinds of trees, even a dense shrub layer, can crowd them out, so that they cannot reproduce and will eventually die out.

The crucial aspect of the environment for longleaf pine communities is fire. We have mentioned the natural occurrence of fire in many of our natural communities, almost always creeping, low-intensity fires that most mature trees easily survive. But nowhere was fire as frequent as in the coastal plain uplands. Abundant thunderstorms and large expanses without natural firebreaks allowed fires to spread into most places as often as every two or three years. Fire coming that often tends to kill most tree seedlings and shrubs and eventually wears out even most of the larger trees. But no tree is as well adapted for frequent fire as longleaf pine. By eliminating other trees and shrubs, fire gives it the elbow room it needs and usually leaves it the sole possessor of the canopy in these communities. At the same time, the open, sunny conditions created by frequent fire make these communities excellent places for herbaceous plants. Longleaf pine communities in natural condition have dense, grassy groundcover that can have many different species.

You can easily see some of the remarkable ways that longleaf pine lives with frequent fire. Instead of growing up into vulnerable miniature pine trees, its seedlings spend their first few years with no trunk. Their most vulnerable part, the bud, stays at the level of the cool ground, while their needles spread upward like a clump of grass. In this "grass stage," fire burns only the easily replaced needles. When the seedlings have stored enough reserves in their roots, they bolt, growing to five feet tall or more in just a year or two. At that height, their crucial bud is above the hottest part of most fires. The unique look of bolting longleaf pine saplings, with their single clump of long needles atop a spindly trunk, is one of the distinctive features of these communities. The graceful look of adult longleaf pine trees, with their open crowns, needles in clusters at the ends of the branches, and lack of lower branches, is also an adaptation to fire. If a rare fire gets hot enough to ignite a branch, the fire won't spread to other branches. And, if a fire gets hot enough to scorch all the needles on a tree, the tree usually just grows new needles and carries on.

The other members of these communities also must be able to live with frequent fire. Wiregrass, like most grasses, readily sprouts after burning and, in fact, grows all the more vigorously. So do all the herbaceous plants in these communities. In fact, many of them won't even flower unless they have been burned. You can always tell when a longleaf pine community has seen fire recently by the flowering stalks of the wiregrass. The shrubs, along with the oaks and other small trees, also easily sprout from their roots if a fire burns them. Most of the larger animals simply move out of the way of fires or hide in underground burrows. Some come right

Prescribed burning is necessary to restore
and maintain fire-dependent communities
in a landscape where allowing natural fire
is no longer a possibility.

back before the smoldering has even stopped to search for prey or seeds to eat amid the ashes. Others return to eat the fresh new growth of the sprouting plants, which appears within a few weeks after a fire. Some of the smaller animals, such as insects that don't fly or burrow, or that exist as eggs or cocoons for a time, are killed by fires. But their huge numbers of offspring enable them to recolonize a burned area as long as there are unburned patches nearby.

Frequent fire maintains the balance in longleaf pine communities, keeping the shrubs and hardwood trees from getting so large or dense that they crowd out the many small plants.

The "grass stage" of longleaf pine seedlings is an adaptation that allows them to survive frequent low-intensity fires that were once common.

With fires no longer roaming freely across the landscape, land managers must replace them with carefully planned prescribed fires to keep these communities ecologically healthy. Ironic though it may seem, only where such careful management is done are longleaf pine communities close to their natural state.

Two animals are particularly characteristic of longleaf pine communities. One is the fox squirrel. These large squirrels are the only ones with the strength to tear apart the stout longleaf pinecones for their seeds. They also help spread the underground fungi that the trees depend on. Fox squirrels prefer to spend most of their time on the ground, an advantage in these open woodlands. The other characteristic animal is the red-cockaded woodpecker. These small (and not very red) birds peck their nest holes out of living longleaf pine trees, usually the oldest, most gnarled ones in the community. Even if you are not a bird expert, you can recognize their presence. Around their nest holes, they peck many small holes which cause sap to escape and cover the tree trunk. Several such cavity trees in a cluster shelter one small family of birds. Because so few of the old longleaf pine trees they need are left, red-cockaded woodpeckers are endangered and have federal protection. The stripes painted on their cavity trees by land managers to warn people not to harm them can help you spot their trees, but the birds themselves are usually out foraging during the day. Both of these characteristic animals need fire as much as the plants do, and disappear as the vegetation structure changes without it.

One unusual kind of dry longleaf pine community deserves special mention. Scattered in the coastal plain are areas of tall, ancient dunes. Though the sand no longer moves, as is the case with the windblown active dunes of the seashore, their coarse, sandy soils have virtually no ability to hold nutrients, and rainfall quickly drains through them. From a plant's point of view, these sand barrens are the driest places in North Carolina, rightly called "deserts in the rain." Longleaf pine and wiregrass can both live in these places, but even they are sparse in them. With them on the blazing, bare sand, you can find a set of specialized plants that would look more at home in Arizona than in most of North Carolina: cactus, yucca, dwarf shrubs with leaves reduced to scales, wiry plants that you can barely see, odd-looking clumps of lichen. The turkey oak, common in most dry longleaf pine communities, also comes into its own in these places. You may notice that all of its leaves are held vertically to the ground; this reduces heating by the sun while letting it use the plentiful light reflected from the sand.

Red-cockaded woodpeckers are rare birds that are strongly tied to longleaf pine. These social birds often forage in family groups.

While most of the vast dry longleaf pine communities are gone, you can still see some of them in the sandhills region and in southeastern North Carolina. You can drive or wander through large expanses in the Sandhills Game Land. You can hike trails through good examples at Weymouth Woods State Natural Area, Carvers Creek State Park, and Carolina Beach State Park. There are also good examples in the southern parts of Croatan National Forest, such as those along the Patsy Pond trail. Large landscapes covered with dry longleaf pine communities still exist on much of Fort Bragg and Camp Lejeune, where accidental fires started by shells and flares have kept them ecologically healthy. The All American Trail, which runs along the border of Fort Bragg, offers glimpses of this landscape. Some of the best sand barrens, the most extremely dry communities, are at Jones Lake State Park and Singletary Lake State Park, though in both cases you need to seek them out in the remote portions of the park. Most of the few remnants of longleaf pine communities in the piedmont region are in Uwharrie National Forest. One of the best is the Pleasant Grove area south of Troy.

Wet Pine Savannas

Like the dry longleaf pine woodlands, wet pine savannas are inviting places, perhaps the most inviting of all of eastern North Carolina's wetlands. If you have seen them, you may not have realized that they were wetlands at all, for they closely resemble the dry longleaf pine communities. Both share an open, airy feeling, with a canopy of widely spaced pine trees, little understory, and dense, grassy groundcover. Longleaf pine is the typical canopy in both, but pond pine, with its shorter needles and knobby cones, may replace it in the wettest savannas. The openness of these communities is accented by the dense pocosin communities that often surround them with a green wall of shrubs. If you visit a wet pine savanna during the growing season, what draws your eye are the flowers amid the grass. Few of our natural communities have as many showy flowers for as much of the year, and none have as many different species. From a few delicate orchids, butterworts, and giant pitcher plant flowers in the spring, the show shifts through the season and reaches a crescendo in the late summer and fall with purple asters, blazing star, coreopsis, sunflowers, grass-of-Parnassus, and dozens of others. Even the flowers of grasses are hard to ignore here, with the lacy purple panicles of muhly and the curved, orange-scented combs of toothache grass calling to be admired. If the obvious beauty does not sate you, you can use a hand lens to admire the feathery bristles of the beaksedges and the surprising textures on the little white seeds of the nutrush sedges that quietly lurk amid the grass and flowers.

Wet pine savannas are a botanist's paradise. They are famous among ecologists for the large number of plant species they support in a small area. Measure out a square meter, barely bigger than a square yard, and comb through it. Beneath the grass are smaller plants and tiny rosettes. An expert can find twenty, sometimes thirty or more kinds of plants in this small area. Lay out a ten-meter square, perhaps the size of your bedroom, and there may be seventy or eighty species. Many of these species occur in no other habitat. Almost no other kind of natural community in the world can match this. Tropical rainforests will harbor more species per acre, or per hundred acres, but not as many in such small areas. Though harder to see,

(above, left to right)
Rosebud orchid is just one of many spectacular orchids found in wet savannas.

The world-famous Venus flytrap can be found only in wet pine savannas of North Carolina and South Carolina.

Rush-featherling is a dominant herb in the distinctive type of wet pine savanna found only in two counties in southeastern North Carolina.

animal diversity likely exceeds that of the plants. Put out a light trap to catch moths flying in a savanna, and you may count a couple dozen to one hundred or more species, just among the larger moths that are relatively easy to identify.

High on the list of botanical wonders of pine savannas are the carnivorous, or insectivorous, plants. No other community harbors as many species equipped to reverse the typical bug-bites-plant story. The tall yellow-green pitcher plants often stand out in savannas, and their shorter red cousins are easy to find if you look amid the grass. The tiny sundews with their knobby hairs and the ordinary-looking but sticky butterworts take a little more looking to spot, unless you are tipped off by their showy flowers in the spring. And our wet pine savannas are the only home for that most famous of carnivorous plants, the Venus flytrap. Indeed, though known to people all over the world, their native range has never extended more than a few miles beyond southeastern North Carolina.

Why should so many plants in these communities have this rare habit? There is a need here. The wet soils of savannas are infertile. Though these plants can make their own food using sunlight, like most plants, they need nitrogen and phosphorus and other nutrients. There is little in the soil here, and plenty in bugs. There is also an opportunity. All the carnivorous plants need plenty of sunlight and can't live in other, more shady communities.

Toothache grass, with its curled, comblike seed heads, is often abundant in wet pine savannas.

Although the carnivorous plants supply their nutritional needs by digesting invertebrates, not all insects are victims. There are specialized moth species that live in pitcher plants, their caterpillars hiding inside the deadly pitchers and then cutting their way out through the side. There is even a Venus flytrap moth, one of the rarest species in North Carolina, whose caterpillars eat only Venus flytraps.

Wet pine savannas, like dry longleaf pine communities, thrive on frequent fire. If you watch a pine savanna closely after a fire, you may wonder how burning can be so important, when its most obvious effects seem to disappear so quickly. A couple of weeks of warm weather are all that are needed after a fire before the savanna is green again. By the fall after a winter or spring fire, the herbaceous plants are at their lushest and most flowery.

The importance of fire can be seen better in its absence. Without burning every few years, the balance of the community breaks down: shrubs become dense; the grasses and other herbs stop flowering and gradually die off; and longleaf pine stops reproducing. After a few decades, these places become such a tangle of evergreen shrubs that only a careful examination of the soil will show that they were once savannas. Even before the shrubs take over, the rich diversity of herbaceous species falls, with the larger grasses and shrubs crowding out the smaller species in just a few years. With the wide-ranging natural fires gone, only savannas managed with careful prescribed burning are ecologically healthy.

As in the dry longleaf pine communities, fire is more of a dilemma for many of the insects. The Venus flytrap moth is a good example. Unable to flee, the moths must recolonize the burned savanna from unburned patches. If the sa-

Common grass-pink is a beautiful and common orchid of wet pine savannas.

vanna burns too frequently or doesn't have enough unburned patches with Venus flytraps in them, the species can't survive there. But any savanna that misses burning for more than a few years will no longer have Venus flytraps for the moths to eat. Working within such paradoxes requires the greatest of skill from fire managers.

Wet pine savannas are among the least obviously wet of our wetland communities. Most are wet because of the high water table in the flat terrain of the outer coastal plain, though similar communities occur in hillside seeps in the sandhills region. But the water table often stays just below the surface even in the winter. If you walk in savannas, you seldom make a splash, and you probably won't sink in their firm sandy or silty soils. But their plants spend much of the year with their roots in waterlogged ground that lacks oxygen. The dominant plants of wet pine savannas, longleaf pine and wiregrass, are equally at home in dry places, but the sedges, carnivorous plants, ferns, and most of the other plants are limited to these wetlands. Conversely, most of the plants of dry longleaf pine communities, from the scrub oaks to the many herbaceous plants in the pea family, are not found here.

Wet pine savannas are not easy to find. Though once common in the outer coastal plain, their easily drained soils and utter dependence on fire mean few have escaped destruction in modern times. Many of the best are hidden away on the state's military bases, in the interior of game lands that are open only during hunting season, or on little-known preserves. The most readily accessible savannas are in the Croatan National Forest, including those along Millis Road and along Little Road. A small savanna can be seen along trails at Carolina Beach State Park. The new Sandy Run State Natural Area, when completed, may offer one of the best opportunities for seeing wet pine savannas.

Coastal Plain Blackwater Rivers and Floodplains

To Americans from other parts of the country, perhaps no natural setting is as evocative of the South as swamps. They conjure up images of mystery or terror: a picture of buttressed trunks rising from black water, beneath the dark shade of a canopy hung with vines and Spanish moss, with snakes and alligators lurking in the shadows. You would not be completely wrong to imagine swamps this way. Such scenes of watery solitude are out there along our blackwater rivers, to be found by those who paddle the back channels at high water. And these rivers are some of the best places in the state to see animals without having to work hard at it. While you are unlikely to see alligators, which remain scarce in North Carolina, any warm day will find many turtles and water snakes sunning themselves on logs by the water. You may see piles of mussel shells left by muskrats, evidence of the aquatic community that is hidden beneath the water. The alarming cry of a pileated woodpecker in the trees or a pair of wood ducks exploding off the water will probably startle you sometime during your visit. A great blue heron, surprisingly large in flight, may lumber from the shore to perch improbably in a tree top. A brilliantly yellow prothonotary warbler may sit in a riverside bush and sing to you. Or you may just drift quietly in the green shadows beneath the tupelo trees and the feathery leaves of the cypress, enjoying as peaceful a scene as nature can muster. North Carolina has a number of different kinds of swamps, but these cypress and tupelo forests of blackwater coastal plain rivers perhaps best fit the expected picture.

If you go searching for such places, you will soon learn that the swamps have other moods as well. Rivers rise and fall, and in dry times that mysterious black water may be confined to the main river channel only. Then the swamp floor may be bare black muck, or it may be covered with lush herbaceous growth: clumps of grassy-looking sedges or the heart-shaped leaves and white drooping spikes of lizard's-tail. One of our most spectacular wildflowers, the spider lily, lights up the early summer swamps with its giant white flowers. Most of the trees and shrubs in blackwater floodplains are deciduous. Even cypress, though it is a conifer, loses its leaves. Visit on a sunny fall day and you will find the canopy glowing with deep red tupelo

leaves or orange-brown cypress needles. A visit in winter will find the trees bare against the sky. In early spring, with the tupelos and cypress being among the last trees to leaf out, the swamps are adorned by the deep red flowers of maples and the yellow jessamine vines in the treetops.

What makes river-floodplain communities different from others? As you might expect, it is the river and its floods. You may be used to thinking of floods as disasters, as they definitely can be for human communities along rivers. But to the natural communities of the floodplain, made up of plants and animals and microbes adapted to water, floods are a beneficial part of life. As in so many natural communities, topography is important, but here, even the topography is created by the river. As rivers meander across the coastal plain, they scour away parts of their banks, and they leave behind sand bars. Old channels, once cut by the river, are slow to fill in after the river changes course, sometimes leaving the floodplain a watery maze. Over time, the bars become ridges of sediment, with swales between them, and sloughs form as the river abandons part of its bed and changes course. Though these landforms may be only a few feet tall, those few feet make a crucial difference. Higher floods will cover even the ridges, excluding upland plants and animals. But the sloughs and swales are inundated by every minor rise of the river, and stay under water for long periods. It is the plants that can tolerate long flooding, like the cypress and tupelo, that thrive there.

Blackwater rivers and streams are those that originate in the flatlands and sandy soils of the coastal plain. If you have seen one of these rivers, you don't have to ask why we call them blackwater. Look into deep water and you see only mysterious darkness. The well-known explanation is that the water is stained with tannins, leached from dead plant material in the swamps. But this is only half of the story. Equally important is what isn't in the water: the suspended clay and silt that are common in piedmont rivers and in coastal plain brownwater rivers. The water is transparent instead of being turbid, and light disappears into its depths rather than being scattered back to the surface. If you look at shallow water with the sun shining through it to a sandy bed, it is a clear red-brown, like tea. This absence of silt and clay is important ecologically. Most river floodplains have very fertile soils because nutrients are deposited by the river with these sediments. But blackwater is acidic and poor in nutrients, and the soils along these rivers are generally sterile sand or organic muck. This affects what kinds of plants can live in the floodplain. The water is also low in oxygen, which limits which fish, mussels, and aquatic insects make up the aquatic community of the river. Both the floodplain community and the aquatic community have relatively few species compared to the communities of other kinds of rivers.

(opposite)
The Black River Preserve, managed by the Nature Conservancy, is the most spectacular old-growth example of what was once a common sight along North Carolina's blackwater rivers. It is home to the oldest stand of trees in eastern North America, with bald cypress exceeding 1,600 years old.

A brilliant yellow prothonotary warbler perches on an overcup oak branch near his nest cavity on the Black River. These colorful birds are common in both blackwater and brownwater floodplains and are usually seen near the river.

Perhaps the most distinctive denizen of the swamps is the cypress tree. Cypress are an ancient line of conifers, whose only living American relatives are the sequoias and redwoods of California. Among their distinctions is great longevity. While most kinds of trees in North Carolina hardly make it past 400 years in the best of circumstances, cypress trees can live well over 1,000 years. North Carolina harbors the oldest stand of trees known east of the Rockies, acres of ragged cypresses that have seen more than 1,600 seasons amid the blackwater. Their ragged and broken tops attest to the many hurricanes they have weathered. But you don't see any uprooted. In fact, it is a challenge to find a blown-down cypress tree anywhere, so firm is their hold on the muck. Cypress trees are long-lived even in death. Their wood is decay resistant, and both ancient hollow logs and the stumps from early logging days persist for decades after other trees have rotted away.

No discussion of cypress trees would be complete without mentioning cypress knees. These knobby wooden protuberances that rise like fantasy towers from the mud around cypress trunks are a sight like no other. They tend to grow up to the level of typical high water,

Coastal Plain Blackwater Rivers and Floodplains

which may be knee-high or over your head. What are they for? An obvious guess in such a wet place is that they carry air to the roots, but experiments to test that idea have come up blank. Perhaps they have something to do with the ability of cypress trees to resist blowing over, but it isn't clear how. Maybe you can think of a different hypothesis. Perhaps they protect the tree trunks from being bashed by floating logs during floods?

Blackwater-river floodplains have other faces besides swamps. Walk or paddle one and you will likely see bottomland hardwoods occupying the higher areas. These forests of wetland oaks are often quite lovely, their wide-spreading branches contrasting with the straight trunks and narrow crowns of the cypress and tupelo. Sometimes the shrub layer is a thicket of the dense, tall, bamboolike stems of switch cane; sometimes it contains the distinctive fronds of dwarf palmetto; sometimes it consists largely of blueberries or other more familiar-looking shrubs. As in the swamps, vines are often prominent. You may also see some of North Carolina's best displays of epiphytes—plants that grow on other plants. The tree branches may not only be hung with shaggy Spanish moss, but covered with beds of resurrection ferns, shriveled in dry spells but open and green as soon as the rain returns.

Sandbars and mud bars may also be prominent if you come at low water. These areas are the youngest children of the river, the recent deposits of sediment washed from other banks and deposited on the inside of bends. Some spectacular wildflowers, like the giant blooms of rose-mallow, may grow in these sunny places.

Blackwater rivers are scattered throughout the coastal plain but are most prominent in the southeastern part of North Carolina. The Black, Lumber, Waccamaw, and Northeast Cape Fear (not the main stem of the Cape Fear) are our largest blackwater rivers, and all offer a lovely day of paddling and exploring. Less well known but also very pleasant are the Scuppernong River, the Cashie River, and numerous other medium-size rivers in the northeast part of the state. If the water is low, you can land on a bar and walk as far as you care to into the swamp or bottomland hardwoods. If the water is high, you can explore the maze of back channels and sloughs by boat. Mosquitoes are often dense inside the swamps in the mid- and late summer but seldom come out over the open river, at least in the daytime. Lumber River State Park offers a rare chance to camp in a floodplain without boating, as well as plenty of places to explore. The tendency of substantial parts of blackwater floodplains to be under water well into the summer means there are few trails in the larger ones. But smaller blackwater floodplains are abundant, and trail crossings of them offer brief glimpses into this watery world. Weymouth Woods State Natural Area and Carvers Creek State Park are two such places.

A spider lily flower brightens a swamp forest on the Black River.

Coastal Plain Brownwater Rivers and Floodplains

Brownwater rivers are the great rivers of eastern North Carolina. Unless you make a hobby of boating or fishing in this part of the state, you may know them only by brief glimpses from highway bridges. But this is a relatively new perspective. Not many generations ago, they *were* the highways. Their deep, muddy expanses of water gliding silently beneath steep banks and towering bluffs were the smoothest route for moving freight and people. Now, as then, most of their shore is a riotous tangle of vegetation: trees of a dozen kinds, thickets of shrubs, canebrakes, draping vines, and lush herbs. Even if you don't know the plants, you can easily tell that there are many different kinds. Peeling white bark, smooth bark covered with warty lumps, tightly ridged pale bark, dark bark, deeply ridged bark all show the diversity of trees, even in winter. Every layer of vegetation is diverse. A casual look may easily reveal eight or more species of vines, from several kinds of prickly greenbriers to grapes to the black-barked, snakelike supplejack.

If you visit in spring or early summer, you will find a vigorous chorus of birdsong filling the air with its tapestry of voices. Squirrels chatter from the treetops. The shadow of a soaring hawk or bald eagle may call your attention even higher. Pushing through the bushes may bring a whiff of the fragrant lemony smell of spicebush to briefly overpower the green scent of leaves and the musty odor of decaying vegetation that fills the forest. If you dodge the greenbriers and watch out for stinging nettles, you may wander for hours through such living exuberance. Or you may unexpectedly find your way blocked by the deep slice of a tributary creek cutting through to the river, its fifteen feet of slippery, vertical bank a barrier nearly as sure as a towering cliff. Wandering away from the river, you may find the land disappearing gradually beneath still waters shaded by cypress and tupelo trees. Here and there you will find mounds of sticks and leaves, piled against trees or logs. These, along with patches of scoured sand, remind you of the flowing waters that recently covered such places and inevitably will again.

Brownwater rivers are the longest coastal plain rivers, entering the region from farther inland. Because they originate in the hilly, clay-rich soils of the piedmont or even the mountains, they carry a load of silt and clay not found in the blackwater rivers. Thanks to the plow and

(opposite)
Flood waters lap at the buttressed base of an American elm tree on a natural levee along the Roanoke River. Gray's sedge dominates a lawnlike herbaceous layer.

114

The Neuse River floodplain in Howell Woods has spectacular examples of mature bottomland hardwood forests including majestic cherrybark oaks.

now the bulldozer, this load has become even heavier in the last two centuries. The water of these rivers is thus a milky red-brown up close. Though their pale look gives less of an impression of inscrutable depths than blackwater does, brownwater is in fact more turbid and harder to see through. When the river floods the land along it, it leaves behind a layer of silt and clay that is rich in nutrients, continually replenishing the soil and making the brownwater floodplains the most fertile places in the coastal plain. The large size of brownwater rivers tends to lead to slower changes in water level; short and sudden floods created by storms can occur, but long-lasting floods driven by long rainy spells are more common.

The landscape of the brownwater-river floodplains is built by the river, even more so than that of the blackwater rivers. With so much sediment to work with, there is plenty of raw material. Though the rivers don't shift around that quickly, over time their meandering indecisiveness has left much evidence of them reinventing themselves. Most characteristic of brownwater rivers are natural levees. When rivers spill out of their banks, the heaviest load of sediment drops right by the river. Over time, the banks have become the highest part of the floodplain—sandy ridges that drop off gradually as you go away from the river. Beyond these ridges often lie large backswamp basins, where floodwaters pool and stand until the clay they are carrying settles out. Old channels and backswamp basins thus gradually fill up, but the deposition doesn't keep up with the building of the levees. Unlike the artificial floodwalls sometimes built along rivers to keep floodwaters from spilling into the floodplain, the natural levees are broken by small, steep-sided channels known as "guts," which allow the rising river to fill the backswamps. Those lower parts of the floodplain then store the water, evening out the extremes of the floods.

River oats is a common grass on natural levees. Its distinctive flowers inspire its other common name, fish-on-a-pole.

Elsewhere away from the river are ridges that were once natural levees, before the channel shifted away. Walk across one of these ridges and it will drop into a swale, but often another ridge will lie a short distance farther. If you fly over the floodplain, or look at an aerial photograph, you can see that these ridges sweep in broad arcs that reflect the meanders of rivers past. As in all floodplains, the height of these landforms determines how long floodwaters cover them, and this determines what plants and other organisms can live there.

The different brownwater-river-floodplain landforms support a mosaic of natural vegetation. Most distinctive are the lush, diverse forests of the natural levees. Their canopies are marked by characteristic trees such as sycamore, sugarberry, box elder, and green ash, shared with the piedmont and mountain floodplains but found almost nowhere else in the state. They join more widespread trees

A barred owl hunts from a river birch branch. The distinctive hoot of this owl is a common sound in river-floodplain communities.

(opposite)
The liverworts clinging to the bottom of these water tupelo in the Meherrin River floodplain indicate that the floodwaters typically rise over six feet deep.

such as sweet-gum, red maple, and wetland oaks. A similarly diverse set of understory trees, shrubs, vines, and even more varied herbaceous plants awaits the careful botanist. As in other diverse natural communities, the herbs spread themselves through the growing season. Spring ephemeral species are abundant on many levees, covering the ground with delicate flowers. A completely different set of larger herbs rises through the summer and into the fall. Often, large, bushy grasses are prominent: river oats with their dangling, flat flowers, or the bottle-brush spikes of wild rye. Sometimes broadleaf herbs such as jewelweed, smartweed, or gold-enrod dominate.

As in the blackwater-river floodplains, bottomland hardwoods and wetter cypress-tupelo swamps are also prominent in the floodplain mosaic. The swamps have the same distinctive, mysterious look of those in the blackwater floodplains, but you might notice that the tupelo trees have larger leaves. This different species, water tupelo, needs the greater fertility found in brownwater floodplains. Bottomland hardwoods reflect the greater fertility by having more species of wetland oaks. In the few mature examples that remain, these trees can form magnificent columns above open ground or shrubby thickets beneath. Often forgotten in the floodplain mosaic, until you run into one, are beaver ponds. The narrow sloughs and guts are easy places for beavers to dam. Open-water ponds, marshy beds of tall sedges or cat-tails, thickets of willows, and stands of young maple trees all join the mosaic as these ponds are created and abandoned. One distinctive thing about coastal plain beaver ponds is that they are not always treeless as those in other parts of the state are. Because the cypress and tupelo trees already growing in the low areas can tolerate long flooding, ponds often have a partial or complete canopy.

Floodplain communities of all kinds are unusually important habitats for wildlife. The fertility and richness of coastal plain brownwater floodplains make them particularly so. Perhaps no place in North Carolina is as alive with bird life in as much variety. Many of the most common birds are shared with the blackwater floodplains: many kinds of warblers, vireos, belted kingfishers, wood ducks, barred owls, pileated woodpeckers, great blue herons, and red-shouldered hawks. But some, such as the Kentucky warbler, the rare cerulean warbler, and the Mississippi kite, are not. Even wild turkeys, at home in most kinds of hardwood forests, are particularly common in brownwater-river floodplains. Mammals too are abundant, though less visible. Black bears, bobcats, and other secretive animals still roam the remote expanses of woods the floodplains provide. As in the blackwater-river floodplains, bats, including several rare species, forage over the open river and roost in the forest.

An often-forgotten aspect of the life of floodplain forests in the coastal plain is that it is partly aquatic. Scoop up a net full of leaves from the floor of a flooded forest and you may find it teeming with aquatic insects and other invertebrates, even tiny fingernail clams. At these times, the forest is the realm of the fish and the duck, more than the bear and the bobcat. Look at a floodplain forest that has been flooded much of the winter and the leaf litter that covered the ground is often gone, not washed away but eaten by things living in the water.

The major brownwater rivers of North Carolina, the Cape Fear, Neuse, Tar, Roanoke, and Meherrin, can be seen by canoeing or motorboating. The smaller ones, such as Fishing Creek, Little Fishing Creek, and Swift Creek, are passable by canoe at some places and times. Trails in brownwater-river floodplains are limited. Vast expanses of protected public lands lie along the Roanoke River, which has been a major focus for land conservation. Extensive opportunities for bushwhacking are present on the Roanoke River National Wildlife Refuge and state game lands. A public trail on the Conine Island unit of the Roanoke River National Wildlife Refuge offers an easy chance to see a bit of our largest brownwater-river floodplain. Medoc Mountain State Park and Cliffs of the Neuse State Park have trails that offer easy glimpses of other brownwater-river floodplains. For a more intense swamp experience, paddle to one of the wooden camping platforms on the Roanoke River for a night.

Wet Forests of Coastal Plain Flats

If you drive through northeastern North Carolina, especially the flatlands around Albemarle or Pamlico Sound, you have seen the subject of this chapter. Dense forests of tupelo, cypress, pine, oak, or dark green Atlantic white cedar trees take their place amid the piney woods, the river floodplains, the tidal swamps and marshes that make up the outer coastal plain. Often they are the backdrop behind fields, but in a few places they line the roads for miles. On a map or aerial photo, they fill many of the blank spots between the rivers and towns, a role they share with the pocosins described in the next chapter. Unless you are a biologist or an avid explorer, though, you probably are not very familiar with these communities. At first glance, they may look so much like the communities of the river floodplains that it takes a while to realize that you are nowhere near a river. But the plants that grow beneath the canopy are thick-leaved evergreen shrubs or other species not typical of floodplains. Venture into the woods and the ground will be damp or sodden, but firm. Rarely can you find enough water to make a splash, and you will not see piles of water-carried debris or other evidence of flowing water. No sloughs or channels will block your wandering, though artificial ditches may. But as in the floodplain communities, if you tune your ears to these sites in the spring and summer, you will find them alive with birdsong. You may come across a deer or even a bear or red wolf, or at least their trails.

These communities are known technically as nonriverine swamps and nonriverine wet hardwoods. The names refer to the fact that these wetlands are not associated with flood-plains, though they share some plants and animals. The fact that there is no more distinctive name for them hints at how unusual they are. Though you can find some elsewhere in North Carolina and in other southern states, the vast expanses in northeastern North Carolina represent the bulk of them. In this low-lying landscape, the last to emerge from beneath the ocean, rivers and creeks are few and far between. While the brownwater- and blackwater-river floodplains are wet because rivers sometimes bring floodwater into them, at other times the same rivers drain water away. Nonriverine flats are wet because the water that arrives in rain-

(opposite)
Laurel oak, with its buttressed base, and swamp chestnut oak, with its pale, shaggy bark, are characteristic trees of nonriverine wet hardwood forests, such as this example at Gum Swamp in Croatan National Forest. Dense beds of ankle-high doghobble, in the background, often cover the forest floor.

A forest of Atlantic white cedar occupies part of the peat-filled Carolina bay at Jones Lake State Park.

fall cannot easily leave. Without creeks to carry it away, it flows very slowly as a thin sheet, or joins the water table that is usually just below the surface. Water never gets very deep on the surface, but the soil stays saturated for much of the year.

As in other wetlands, the saturated soil requires special adaptations by plants and other organisms that live in the soil. Oxygen is scarce around the roots. Few upland species can survive these conditions. But there are plenty of wetland species ready to take advantage of them, and a wide variety of natural communities develop in response to variations in the environment. Nonriverine flat wetlands are home to the wet pine savannas and to pocosins, the subject of the next chapter, as well as supporting the variety of dense swamps and wet hardwood forests we are discussing here.

The most abundant remaining nonriverine wetland forests are the swamps dominated by tupelo and cypress. These are the same trees that dominate the wettest parts of blackwater-river floodplains, but the understory, shrub, and herbaceous layers are quite different. Many of the larger animals are the same, though the fish and aquatic invertebrates that inhabit deeply flooded river swamps are absent.

In less wet flats, the nonriverine wet hardwood forests contain trees similar to those in the bottomland hardwoods of brownwater rivers: laurel oak, swamp chestnut oak, cherrybark oak, sweet-gum, and others. But beneath them will be red bay, doghobble, large ferns, or other plants tolerant of infertile, saturated soils. These wet hardwood forests were once common, but, being less wet than the swamps, were easier to drain and convert to other uses. They are now one of the most endangered community types in North Carolina. Most endangered of all is the wet hardwood forest on shallow limestone: the only example in the world is found near Rocky Point, in southeastern North Carolina.

One other kind of forest of wet flats that deserves special mention is the Atlantic white cedar forest. Atlantic white cedar is a distinctive tree, with its soft, vertically striped bark, dark green scaly foliage, narrow conical crown, and characteristic smell. Visiting old Atlantic white cedar forests is an experience like no other, one that few people have had. The cedar wood is very resistant to decay, and the new trees grow right on top of the logs left by their predecessors. There is sometimes nothing you can recognize as a ground surface to walk on, only a tangle of logs, the spaces between them loosely filled with dead cedar needles. Every step is a balancing act.

Atlantic white cedar is also distinctive in its ecology. Most of our forest communities, if left in their natural state, have trees of all ages mixed together. New trees grow up in gaps where individual trees or small groups have died, a few at a time. But Atlantic white cedar forests naturally tend to occur as even-aged stands, with all trees having gotten their start about the same time. Under natural conditions, large expanses of these relatively short-lived trees were killed by fire or toppled by windstorms, allowing subsequent mass reproduction to occur. More recently, they sometimes regenerate in a similar way after clear-cutting, but more often other trees regenerate and the community is lost. Logging and lack of fire have made Atlantic white cedar forests rare, though they once were very common in northeastern North Carolina.

Wet forests of coastal plain flats are present on several public lands, but access is not easy for most. Dismal Swamp State Natural Area is one place where you can see some nonriverine swamps along trails. Trails just over the Virginia state line in the Great Dismal Swamp National Wildlife Refuge offer other glimpses. You can see some of the most extensive remnants of nonriverine swamps along the roads in Alligator River and Pocosin Lakes National Wildlife Refuges, and in Buckridge Coastal Reserve. Some of the best Atlantic white cedar forest remnants are in remote portions of Alligator River National Wildlife Refuge, but you can see a small example along a trail in Jones Lake State Park. The one readily visited nonriverine wet hardwood forest is along Pine Grove Road in Croatan National Forest, where you can recognize it by the swamp chestnut oaks visible along the road. Though privately owned and not publicly accessible, the one remaining example of the limestone wet hardwood forest can be glimpsed from I-40 near the town of Rocky Point. It can be recognized by the abundance of dwarf palmetto beneath the hardwood canopy.

Rocky Point Marl Forest contains a distinctive type of wet hardwood forest found nowhere else in the world. Dwarf palmetto forms a dense shrub layer in much of this community.

Pocosins

If you have hiked, explored, or even driven in the coastal plain, you have probably seen pocosins lining the roads or forming a backdrop to other communities. But you may not have realized that the wall of brush and sparse, crooked pine trees was a natural community at all. At first glance, they don't look that different from the brushy expanses of sparse, small, crooked trees that are often left by logging operations. If you walk along the edge and look more closely though, you might appreciate how picturesque some of those pines can be—old-looking, gnarled graduates of the school of hard knocks, or odd bottlebrushes with pine needles growing right out of branchless trunks. Come in the right season, and you can see thousands of little bell-like flowers decking the bushes, or big, showy white flowers on the scattered hardwood trees. You might notice the quiet background hum of countless bees as they visit bushes with fragrant flowers so small you hardly notice them. Or maybe the varied song of a catbird will reach your ears as it sings from the tiptop of a tall bush. If you pass by such a place frequently, you might be struck by how little it seems to change with the seasons, for few communities are as evergreen. Virtually all the large plants hold their leaves all year—the hardwoods, the bushes, even the vines, as well as the pine trees.

Perhaps the most common thought people have about pocosins is, "I'm glad I don't have to go in there." Few indeed are the people who have ventured far into a trackless pocosin. These sites are perhaps the least user-friendly of all our communities. The inside experience is one of swimming or wrestling through the dense, intertwined shrub stems and branches. The dead leaves and small twigs that you break off as you push through the brush find their way into your clothing and accumulate in little piles in every pocket. You must keep a careful eye out for the long spikes of greenbrier stems, or pay for it in skin. The ground surface is sometimes only a general concept rather than a firm reality, since it progresses gradually from live shrubs to dead sticks and leaves to soft muck. While one step may be onto a buried log, the next may find your foot plunging into a black soup that seems to have no bottom. If you find a trail to follow, the shrubs are less of a barrier, but the soft muck holes are more frequent and deeper. The most inviting open spots are sometimes the least feasible for walking.

If you persevere through the tallest, thickest vegetation on the edges, you may reach an area of shorter shrubs, where you can wade rather than swim through the vegetation. There the pocosin may be a maze of little pockets of short vegetation walled off by partitions of taller shrubs, or it may be a sweep of evergreen sea punctuated with gnarled pines beneath the open sky. Not all pocosins have such an open center, and some such openings are too far to reach from the edge. But if you spend a day pocosin crashing, though you may end it feeling like you've climbed a high mountain and wrestled a bear, you will have experienced one of the wildest places in eastern North Carolina.

Pocosins are often called evergreen shrub bogs. And it is indeed the shrubs and the wet organic soils that make pocosins what they are. The characteristic tree, pond pine, may occur as a fairly dense canopy or as only a sparse scattering of stunted individuals, but the shrubs are always dense. A small set of typical species makes up pocosins, generally fetterbush, titi, gall-berry holly, honey-cup, red bay, and loblolly bay. Most of them have evergreen leaves that are thick and of a nondescript oval shape. Also pretty much constant is laurel greenbrier, some-times known as bamboo-vine for its thick dark stems, or blaspheme-vine for the response it elicits from the traveler. It too possesses thick, oval, evergreen leaves, along with stout prickles on its stem. One researcher described it as the plant that holds the pocosins together, a claim that is both figurative and literal. Pocosins have few herbaceous plants, but among them are the distinctive forms of pitcher plants and sundews, and the bladderworts whose little showy flowers seem to float above the puddles in the spring. Pocosins share with the wet pine savannas the distinction of being primary habitats for carnivorous plants. Indeed, a larger percentage of their plant species catch and digest bugs than any other natural community.

Pocosins are kind of a North Carolina specialty. While they occur in neighboring states, North Carolina has the lion's share in terms of acreage and variety. The name pocosin is said to derive from an Algonquian Indian word meaning "swamp on a hill." While you may find the idea of a hill in the flat outer coastal plain a bit odd, the larger pocosins are indeed on the broad areas farthest from the rivers. Maps sometimes show that streams flow out from the large pocosins in all directions, demonstrating that they really are gentle hills that are the highest parts of their landscape. Other pocosins fill the mysterious oval-shaped depressions known as Carolina bays, or lie in smaller swales and basins. In contrast, in the sandhills region, pocosins tend to lie at the bottoms of the hills, along the small creeks and shallow

(opposite)
Gnarled and stunted pond pine trees stand amid dense shrubs in the Pocosin Wilderness, Croatan National Forest.

Sweetbay magnolia, with its large showy flowers, is one of several small evergreen trees that occur in pocosins.

Laurel greenbrier, the plant that holds the pocosins together.

ravines. What these places have in common is that the soil is saturated nearly all the time but never really floods. The water comes directly from rainfall or seeps out of ancient sand dunes. It carries almost no nutrients, making these places perhaps the least fertile in the state.

Pocosins have organic soils, known as peat or muck. Most places have a layer of fallen leaves and sticks on the surface, which nourishes the soil as the organic materials decompose. But beneath this surface, the soil is made up primarily of sand, silt, or clay. In pocosins and some other wetlands that are constantly saturated, the organic material can't fully decompose. Lack of oxygen in the saturated soil inhibits bacteria and fungi, and the plants that grow there produce leaves that are particularly hard to break down. Without decomposition, nutrients aren't released from dead material, and infertile soils become even less fertile. The organic soil also acts as a sponge, raising the water table and making wet places even wetter. Over the thousands of years since the last ice age, the organic layer has piled up ten feet thick or more in many pocosins. In some places, it has spread across the landscape to cover many square miles.

Besides wet, organic, infertile soils, the crucial aspect of pocosin ecology is fire. Most of North Carolina's communities naturally burned, and we have mentioned the important way fire sustains some of them. The behavior of fire in pocosins is particularly dramatic for its intensity, its unpredictable behavior, and its paradoxical effects. To people whose job it is either to put out wildfires or to replace them with prescribed fires, pocosins are among the most difficult communities to work with.

Being wet as sponges and full of green leaves all year, pocosins would seem unlikely to burn at all. And indeed, most of the time they won't. Land managers often conduct prescribed burns in pine savanna communities using the bordering pocosins as fire breaks. But in the spring, when the new leaves look the freshest and wettest, the vegetation is volatile and much more likely to burn. And during droughts, all bets are off. Most of the large "forest fires" that have happened in eastern North Carolina in recent years have actually burned in pocosins. With the right wind, fearsome, fast-moving pocosin fires can easily jump roads, creeks, and firebreaks, spreading over thousands of acres in a day. If the fire occurs in a drought, the organic soil itself can ignite. Once burning, these peat fires can smolder for weeks. Some have burned all summer, or have been quenched only by the torrential rains of tropical storms.

The immediate aftermath of one of these fires is dramatic in its own way. The recent sea of green is a jungle of black stems or a stubble of small black stumps. The pond pines are stripped of their needles, and often their branches as well, so that only charred black trunks remain above the black ground. A deader scene is hard to imagine. But just a few months later, sometimes just a few weeks, fresh green leaves, twigs, and briers pop up from roots and tubers. New pine needles sprout right from the trunks of scorched pond pines, eventually followed by new branches. The tight little cones of the pond pines are triggered by the heat to open and drop their seeds, so a new generation of pines can get a start on the bare ground. Fed by all the nutrients released by the fire, vegetation grows rapidly. If you want to take advantage of the opportunity to walk unimpeded through a pocosin, you must not delay. While peat with artificial ditching may burn deeply enough to kill the roots of the plants, in natural pocosins the vegetation will generally be as tall and dense as before in just a couple years. Thanks to an amazing set of well-adapted plants, the most dramatic fires in North Carolina can have little lasting effect.

While the interior of most large pocosins is extremely hard to visit, one place to get a feel for them is the center of Croatan National Forest, where Catfish Lake Road crosses a large peatland. Stop at the road to Catfish Lake to see the vegetation close at hand. Highway 264 in Dare County runs for miles through taller pocosins on the Alligator River National Wildlife Refuge. Other large pocosins occur on Holly Shelter Game Land, Angola Bay Game Land, and Pocosin Lakes National Wildlife Refuge, as well as Croatan National Forest. Fairly large pocosins occur in many of the Carolina bays of Bladen Lakes State Forest, Jones Lake State Park, and Singletary Lake State Park. The park trail around Jones Lake offers a rare opportunity to see one from the inside, as does a canoe on the lake. Smaller patches of pocosin are present in Carolina Beach State Park, Weymouth Woods State Natural Area, Sandhills Game Land, and Boiling Spring Lakes Preserve.

The palamedes swallowtail is one of the most common butterflies of pocosins, but the caterpillars depend on red bay, which may be threatened by a new disease spread by an exotic insect.

Coastal Plain Small-Basin Communities

Coastal plain small-basin communities can be like secret jewels set in the expanse of the piney uplands. A polka-dot fabric of lily pads on sparkling water shines in a sudden break in the trees. A brilliant yellow or purple carpet of flowers glows amid the grass. A brooding grove of cypress trees with swollen bases offers its counterpoint to the columnar trunks of surrounding pines. Other basins are more subtle, with only a different kind of grass to mark them. These may be hardly noticeable in the summer, though they brim with water in the winter. Some ponds beckon you from a hilltop, inviting you to come sit on the shore and watch the dragonflies dart and hover. Some call you from afar with a chorus of frogs amid the whispering pines. Others are surrounded by a wall of dense shrubs and vines, so that you must brave greenbriers or wrestle the interlocking branches to see what surprise awaits within. Most are small, but some are large enough that you can lose yourself in a world of cypress trees and water, grass and flowers.

These communities inhabit shallow depressions in the coastal plain landscape, where water stands for at least parts of typical years. In most basins, water levels fluctuate with the seasons. Shallower ones go from flooded to dry in the course of a typical year. Even the deeper, permanent ponds can have a broad rim that is submerged for only part of the year. Water levels fluctuate in longer cycles as well, with some basins remaining dry all year in droughts and staying flooded all year in wet periods. Some coastal plain basins are Carolina bays: mysterious ovals, all lined up the same direction. Some are limesinks, where buried limestone has dissolved, causing the sandy surface to collapse. Some are deep swales in sand dunes.

If you visit several of these communities, you will quickly notice how variable they are. Many depressions are strongly zoned, with rings or patches of vegetation such as a dense shrubby rim, a diverse meadow in the sandy edge that dries out early, a tall marsh in deeper water, and water lilies in the permanently flooded center. Trees, usually pond cypress or swamp tupelo, are present in some depressions, or some zones of a depression, but not in others in the same area. Flat-bottomed basins such as Carolina bays may not be zoned at all but may have uniform vegetation all the way across.

Coastal plain small-basin communities hold great botanical interest. Wetter zones tend to have a few species of grasses, sedges, or water lilies that dominate large areas, and drier zones tend to be dominated by shrubs and trees, but both include some species you won't find anywhere else. The drawdown zones, those areas that flood but dry out early in the season, usually support the greatest diversity of plants. Some of these plants are very showy, such as the tall pinebarren milkwort, with its big, flat-topped, yellow flower clusters, or the several meadow-beauty species with their big purple flowers. Others are picturesque in their oddity, such as the fuzzy white knobs of bogbuttons or the scaly cones and tiny yellow flowers of yellow-eyed grass. And even more diversity lies hidden amid the drab-looking grasses and sedges.

If you are looking for rare plants, these communities are among the best in the state to search. Half a dozen or more rare plant species often occur in a single pond. They range from tiny sedges only an expert would notice to spectacular wildflowers such as the awned meadow-beauty. Many of the rare plants in these communities follow an interesting geographic pattern. They have wide ranges, occurring across many states throughout the coastal plain. But they are everywhere scarce, with populations few and far between. It is often suggested that seeds may hitch a ride on the feet of migrating waterfowl from one patch of this rare habitat to another.

If you studied a given pond closely over many years, you might notice another of the distinctive ecological features of these communities. The vegetation of many of them can vary sub-stantially from one year to the next in response to variations in wetness. Once-abundant plants may disappear without a trace, while plants you haven't seen in years may suddenly carpet the basin. This is made possible by a phenomenon known as seed banking, whereby some seeds produced by a plant stay dormant in the soil and don't germinate until a later year. It takes a number of special adaptations for seeds to do this. They must refrain from germinating right away; stay alive for years amid wetness, fungi, bacteria, and hungry animals; and then recognize the right time to germinate later. Like money in the bank, it is a good adaptation for plants whose habitat may not always be habitable to them. Seed banking is common among desert plants but is much less common in North Carolina. Studies have shown that many coastal plain small-basin communities are champions at seed banking, with large numbers of their plants able to bide their time in this way.

Like the piedmont upland swamps and pools, coastal plain small-basin communities are particularly important to amphibians. If you visit one in

Flowers of the rare pondberry, a relative of the more common spicebush, in a seasonal pool in a Carolina bay in Sampson County. This species has only four remaining occurrences in North Carolina and is rare throughout its range.

the early spring, especially at night, it may be alive with the chorus of frogs. These sites are the most important breeding habitats for some of the rarest amphibians in North Carolina, such as the Carolina gopher frog, ornate chorus frog, dwarf salamander, and tiger salamander. They are important for common species as well and can be notable for their diversity of amphibians. An expert herpetologist once recorded seventeen species of frogs in a single year at a single coastal plain depression, a record unlikely to be broken.

As with all communities that occur as small patches, the surrounding landscape is of great importance to these communities. As in the piedmont upland swamps and pools, the seasonal convergence of amphibians on these fish-free ponds is the most dramatic illustration of this. But a look at the muddy tracks of raccoons and other upland mammals on the edges of ponds illustrates the kind of connection that is always present. So too do the dragonflies that dart through the nearby upland forests.

Coastal plain small-basin communities are rare, but there are several places where you can easily see them. Carolina Beach State Park has many ponds and is a great place to see their variability. Patsy Pond, in the southern part of Croatan National Forest, has a public trail that leads visitors to several spectacular ponds. Some of the even-rarer examples in dune swales can be seen along trails at Buxton Woods in Cape Hatteras National Seashore. Very different examples can be seen at the Nature Conservancy's Nags Head Woods Preserve.

(opposite)
White water-lily flowers in the Patsy Pond
area of the Croatan National Forest.

Natural-Lake Communities

North Carolina's natural lakes are among the easiest of coastal plain places to love, offering an airy openness amid the dense pocosins or deep forests that surround them. Dancing ripples catch the sun as a welcome breeze plays across the water. Picturesque cypress trees line the shallows, artfully spaced amid grass beds or clumps of bushes. If you wade out into the warm summer waters—far out because the slope of our lake beds is very gentle—you can see the sunlight break into abstract shafts and streamers that play on the sandy bottom. Or you may come in winter to view the teeming flocks of tundra swans, geese, and other waterfowl that make some of our lakes their seasonal home.

Though much larger than the small-basin communities, most of our natural lakes offer a kind of intimacy. All have rounded shapes, lacking the coves and arms of artificial lakes. You can see almost all of the lake from any point. At the same time, where houses or clearings don't line them, a dense wall of shrubs or trees usually keeps these lakes remote from the surrounding world. But you may find the largest of our lakes, such as Lake Waccamaw and Lake Phelps, a different experience. The opposite shore, several miles away rather than the more typical one mile, seems remote. With more room to work, the wind can seem more serious than playful. Here, substantial waves can pound the shore, and you are wise to think about the weather before venturing out in a small boat.

Natural lakes are rare in North Carolina. There are fewer than twenty, all in the coastal plain. But this is a wealth of them compared to most of the southeastern United States. Virginia has only two, South Carolina none. Only Florida has more. Though we often don't appreciate the fact, lakes are a geological oddity. The lifespan of a lake is an eyeblink compared to the life of a mountain or a river. Sediment fills basins, or outlet channels erode to drain away the water. Without the glaciers that created most of the lakes in the northern and western United States, North Carolina's lakes are perhaps even more of a geological oddity. Many lie in the mysterious oval-shaped basins of Carolina bays. Most of the rest are embedded in large peatlands, where they may, or may not, have been excavated by enormous, deep peat fires in the dim past. While many artificial lakes have been built in North Carolina, from large reservoirs to

(opposite)
Tundra swans rest on Lake Phelps
on a winter morning.

Cypress trees and maidencane line the shore of Lake Waccamaw.

small millponds and farm ponds, none is similar to the natural lakes in form, in inhabitants, or in having a character shaped by thousands of years of natural development.

Most of our natural lakes have clear or tannin-stained, acidic water, similar to the blackwater rivers. Most don't have a stream that feeds into them, so water comes from rain or filters in through the surrounding peat. These acidic waters, naturally low in nutrients, support the relatively few species of fish and aquatic insects that do well in these conditions. The vegetation that occupies the shoreline zone is picturesque, but similarly has few species. Pond cypress, maidencane, and a few kinds of shrubs prevail, and there is nothing like the flowery meadows of many small-basin ponds.

One lake is an exception to most of these patterns. Lake Waccamaw is fed by seepage from sands that contain calcium, and a limestone bluff lines part of its shore. As a result, its clear, shallow waters are not acidic. Amid the pond cypress and maidencane, there are numerous smaller herbaceous plants. Beds of the long, crenellated, floating leaves of narrowleaf pondlily wave in the water.

Even greater biological treasures lie hidden beneath Lake Waccamaw. Its sandy bed is packed with mussels. These mussels belong to many different species, and several of them are endemic to Lake Waccamaw—they occur nowhere else in the world. There is an endemic snail species as well. The little fish that swim around your wading feet may well also be of a species found only in this lake. You would have to journey far to find another lake with so many species whose whole existence is confined to so small an area. No other coastal plain lake with abundant calcium exists north of Florida. The calcium is especially important for the mussels, which need it to build their shells. Some biologists joke that, if you took a Lake Waccamaw mussel and put it in another lake, it would dissolve. That may not be literally true, but it would be no laughing matter to the mussel. For these special species, Lake Waccamaw offers a home that no other lake can match.

Despite their rarity, there are many opportunities to visit North Carolina's natural lakes. Though houses lining the north and west shore of Lake Waccamaw limit access there to owners and renters, Lake Waccamaw State Park gives you a place to experience its vastness and see its distinctive shoreline community. You can also launch a boat from a Wildlife Resources Commission boat ramp, but beware of the wind and waves that often develop in the afternoon. Jones Lake and Singletary Lake State Parks offer access to lovely small lakes, while Pettigrew State Park provides a chance to see much larger Lake Phelps. Croatan National Forest offers rustic access to Catfish Lake and Great Lake. Lake Mattamuskeet National Wildlife Refuge's wildlife drive and overlooks are rightly famous among birders for their splendid opportunities to see the vast flocks of wintering waterfowl. Pungo Lake in Pocosin Lakes National Wildlife Refuge offers a less-well-known lake and wildlife experience. Milltail Lake in Alligator River National Wildlife Refuge and Hidden Lake in the Conservation Fund's Palmetto-Peartree Preserve offer a chance for a quiet paddling trip on lakes that have a different feel from most others.

Narrowleaf pondlily leaves float in the shallow water of Lake Waccamaw.

Tidal Freshwater Swamps and Marshes

If you travel down one of our coastal plain rivers, whether blackwater or brownwater, you will notice gradual changes as you move toward the coast. The banks become lower. If you are looking for a dry natural levee to camp on, you find yourself worrying as they appear less frequently and are smaller. The wet swamps of tupelo and cypress, confined to sloughs and backswamps upstream, become common along the widening river until you see nothing else. Deep, wide back channels branch from the main river, quiet and still as lakes. Take your boat up one of these back channels and you may reach a mile or more back into the swamp before it ends in a splay of dead-end creeks amid the Spanish moss drapery.

You may notice a change in the behavior of the water as well. While the current gradually became slower as you went downstream, here it may stop altogether, maybe even flow upstream a bit at times. If you stay in one place for hours or days, you will see the water rise and fall. Terra firma in the morning may be flooded in the afternoon and back to a foot or two above the water again by evening. But, you may find that terra firma is not firm at all. The soil of these communities sometimes is soft muck which will not support your weight, and trying to walk in them can be a messy and somewhat-risky proposition.

As you continue downstream, beds of tall grass appear along the banks of the wide river, first in small patches, then large. The swamps may continue as dense, dark, mysterious forests, but you may instead find that the canopy is open. With light pouring through the sparser overhead leaves, lush herbaceous and shrub growth blankets the ground. You may recognize the lizard's-tail and smartweed of the swamps upstream, but here they keep company with many newcomers. Look closely and you will be amazed at the variety of plants that occupy these open woodlands. In summer, there are showy beauties such as the hanging orange flowers of jewelweed, three-petaled white arrowhead flowers, the purple spikes of pickerelweed, the neat pink flowers of swamp rose. There may be a dozen different forms of the green, spiky flowers of sedges alone. While the warblers, wood ducks, barred owls, and many other birds of the floodplain forests upstream have followed you down this far, ospreys may be more abundant.

Your tour into freshwater tidal communities may end here, with further travel bringing you into salty waters and into communities we will discuss in the next chapter. But in a few places, such as the Cape Fear River or the rivers that feed Currituck Sound, your freshwater journey can continue into wetter areas. Here you see the open swamp canopy give way to dead snags, then to large expanses of open marsh. Unlike the brackish and salt marshes of the next chapter, with their uniform vegetation dominated by just one plant species, these marshes are patchy and quite variable. One patch may be sawgrass, one needle rush, others giant cordgrass, cattail, or three-square sedges, all standing above your head. Under them may be arrowhead, green arrow-arum, pickerelweed, and a host of other broadleaf herbs. Here the forest birds have stayed behind, and red-winged blackbirds and marsh wrens perch on the tall grasses and sedges while ospreys soar overhead. Other animals haven't changed though. You may still see a muskrat swimming by, turtles sunning on logs along the bank, or water snakes draped in bushes.

Freshwater tidal swamps and marshes are among the wettest of our wetlands. Sitting right at sea level, they never dry out. Tides may cover them with just a few inches of water or with several feet, twice a day or at more irregular intervals. Even when the tide is down, the soil remains saturated and low in oxygen. Often the soil is organic. The soils in these communities are not as infertile as those in pocosins, however, because the tidal flooding brings in nutrients.

You may already be wondering about the paradox of freshwater tidal wetlands. The ocean moves with the pull of the sun and moon, and its water moves in and out of the sounds and river mouths as it rises and falls. But the ocean is salty. How can we have tides in fresh water? There are two ways. On rivers such as the Cape Fear, the amount of freshwater flowing down the river limits how far the seawater can penetrate upstream at high tide. The rising ocean may actually push the fresh river water back upstream, or it may simply act as a dam and cause the river's water to pile up behind it. Either way, the river rises and falls twice a day like the ocean. The other mechanism, called a wind tide, can occur in the freshwater and brackish sounds where there is little connection to the ocean. Though wind tides are irregular and not as frequent as twice a day, they are influential enough to create similar natural communities. We will talk more about wind tides in the next chapter.

Because they occur right at sea level and because their patterns are closely tied to wetness and the amount of salt in the water, freshwater tidal wetlands

Sawgrass is a common dominant plant in tidal freshwater marshes, often forming dense, tall beds. The fine, sharp teeth along the leaf margin give this plant its name.

A wind-tidal swamp of cypress and tupelo lines the Chowan River.

are the best place to observe the effect of changes in sea level. While the warming oceans and melting ice caps can be expected to increase the rise in sea level in future years, the water has already been rising gradually for years. You may have noticed, in your journey downstream through the tidal swamps, that you came to a place where the canopy thinned, and then perhaps to an area of marsh with dead trees. These are places where tidal swamps once flourished, but where the increased wetness first stressed the trees and eventually killed them. The plants and animals of the marshes start moving in as soon as the canopy opens and full sunlight reaches the ground, so that by the time the trees finally die, a ready-made marsh usually is present below. If you could watch the changes over many years, you would see the freshwater marshes marching upstream and inland, the tidal swamps giving way before them. Marching is a better metaphor than creeping, for the progress is not necessarily steady. Storm surges or extra-high tides can be the "steps" that carry the marsh inland a long distance all at once.

Though it is less obvious, the same thing is happening at the upper edge of the tidal swamps. As rising sea level lets tidal waters penetrate farther upriver, and as the soil becomes more permanently saturated, the river swamps upstream gradually turn into tidal swamps. The same happens inland of the low sound shorelines, where nonriverine wet flats lie behind the tidal swamps. Because cypress and tupelo form most of the canopy in all these communities, you may not notice much change. But any associated pines, Atlantic white cedar, or other less water-tolerant trees will die, and the crowns of the tupelos and maples will grow thin as the wetness increases. The understory, shrub, and herb layers will change. The shrubs of the nonriverine communities, like fetterbush and gallberry, will die, and wax myrtle and swamp rose will replace them. Chain ferns will give way to the diverse set of herbaceous plants of the tidal swamps.

While this process of community adaptation and change is natural and relatively orderly, there is an artificial version of it that is a serious problem. In the wet flats and pocosins of northeastern North Carolina, many miles of ditches and canals were dug to get fill material for roads and to drain away water. As sea level rises, storm surges and then wind tides reverse that role, bringing deeper and saltier water far inland from the natural tidal wetlands. This can kill large areas of swamp and pocosin without allowing natural tidal communities to develop.

Osprey are common in freshwater tidal wetlands.

(opposite)
Royal fern and lizard's-tail grow in the soft muck beneath water tupelo and bald cypress in a tidal swamp along the Chowan River.

Restoration of natural water flow by blocking the ditches, or simply installing one-way gates to keep out tidal water, can help prevent this destructive, too-rapid shift in vegetation.

Freshwater tidal swamps and marshes are easiest to visit by boat, but you can get glimpses from highway crossings and boat ramps in many places. Given the constant wetness and frequent soft ground, virtually no trails exist in them. One rare opportunity to walk through one is the boardwalk at the visitor center in Columbia, North Carolina. Miles of tidal swamp, much of it protected as state game lands, line the Chowan, Alligator, North, Northwest, Pungo, and Cashie Rivers, as well as the shorelines of Albemarle and Pamlico Sounds and their many smaller tributary creeks. Some are also present at the mouth of the Roanoke River and Tar River. These are all wind-tidal swamps. Extensive regularly flooded tidal swamps line the lower Cape Fear, North-east Cape Fear, and Black Rivers, as well as the lower Neuse, White Oak, and New Rivers. You can see some of our most extensive freshwater tidal marsh where highways cross the Cape Fear River at Eagle's Island, opposite Wilmington. Our other extensive freshwater tidal marshes line Currituck Sound and its tributaries, especially the Northwest River and North River. You can most readily view some at Mackay Island National Wildlife Refuge on Knotts Island. Large expanses protected as game lands along the Northwest River, North Landing River, and North River can be visited by boat.

(opposite)
Bald cypress are faring better with rising sea levels than the other trees near the mouth of the Scuppernong River.

Salt Marshes and Sounds

If you continue the downstream journey we began in the last chapter, you will eventually pass through the tidal swamps and freshwater marshes and find yourself in brackish or salty water. Here is a world of open expanses that contrast with the closeness of most of our natural communities. Vast sweeps of marsh, broad coastal rivers, wide bays, and sounds where the water stretches into the wave-rippled distance await you. Unlike the intricate mosaic patterns of the tidal freshwater marshes or the layered forest structure of the swamps, the marshes here are painted in broad brushstrokes. Uniform-looking patches in the yellow-green tint of cordgrass or the blackish-gray of needle rush can stretch on for acres, even for square miles. Not all of these communities are huge, though. You can find small patches along the myriad little tidal creeks and bays that line the sounds.

These watery places are estuaries—places where freshwater and ocean meet, in bays and sounds and the mouths of rivers. The ebb and flow of tides, the persistent push of wind, the concentration or dilution of saltwater, rule the environment for the organisms that live there. Though life on earth originated in the salty waters of the ocean, salt is harmful to plants and animals that have adapted to life on land or in freshwater. At the same time, organisms adapted to the salinity of the ocean suffer when its salt is diluted. The number of species that thrive in the intermediate and varying salinity of estuaries is relatively small. But for those that can tolerate it, these communities are a fertile field with moderate temperatures and an abundant supply of nutrients and water. Healthy estuaries are among the most productive of ecosystems, producing plant material that feeds large populations of fish and shellfish, exporting food to the ocean, and piling up organic sediments.

One distinctive characteristic of some of North Carolina's estuaries is the importance of wind tides. Waters in the southeastern part of the state, north to around the Neuse River, are well connected to the ocean and rise and fall regularly with the ocean's response to the moon and sun, as is typical of estuaries in most parts of the world. You might think this would necessarily be true of any place that is connected to the ocean enough that the water is brackish. But

(opposite)
Saltmarsh cordgrass dominates the marsh at
Cedar Point in the Croatan National Forest.

northeastern North Carolina is different. The tidal range in the ocean here is very small, and tidal inlets are few compared to the large extent of the sounds. Without the steady rhythm of the sea to drive them, the sounds are subject to the vagaries of wind. A strong wind blowing steadily across the large expanses of a sound actually pushes the water to one side. A nor'easter, blowing in from the coast, can raise the water on the inland side by several feet, flooding both brackish and freshwater marshes. The northwest wind that follows a spring cold front can push the water out of the sounds and drop the inland portions by a couple of feet. The result is tidal flooding that, though irregular, is enough to create marshes barely distinguishable from those that are more regularly flooded by ocean tides. Northeastern North Carolina is distinctive, if not unique, in having so much wind-tidal water and marsh.

Salt concentration is the most crucial aspect of the environment that creates differences in these communities, and the differences are easy to see. Where the water is close to full-strength seawater, near the ocean inlets, saltmarsh cordgrass dominates the marshes. Its relatively broad leaves are a distinctive, attractive yellow-green. Where the water is brackish, diluted but still salty, black needle rush takes over. A casual walk though a needle-rush marsh will show you that this plant is well named. The round stalks that serve as both stem and leaf for the rushes are each tipped with a sharp point, making forays into brackish marshes a prickly experience. Besides the dominant grass or rush, you will see few other plants. These communities are among the least diverse of any when it comes to plants. The few that they have can be showy, including the giant white or pink flowers of the bushy marsh-mallows and the yellow spikes of specialized goldenrod species with succulent leaves. But only on the narrow upper edge of the marsh, where just the highest of tides bring salty water, do you see more kinds of plants.

The animals of the marsh community are similarly limited in diversity but may be quite fecund. Animals are easier to see here than in most communities. At low tide, you can see thousands of periwinkles—large snails—perched on the marsh plants. Crabs scuttle along the exposed ground by the dozens or hundreds. Herons and egrets stalk the shallow waters, while cormorants and ospreys soar overhead and ducks paddle the open waters. Secretive rails may lurk in the dense vegetation, but only the most patient, or lucky, birder is likely to see them.

Though you can't generally see it, there is plenty of activity under the creeks and bays, as well. In contrast to the communities on land, the plant component is not the largest and most visible in most aquatic communities. The plants are mostly single-celled microscopic algae, tiny but numerous and fast-growing enough to feed the animals. Crabs, horseshoe crabs, beds of oysters, and the young of oceangoing, as well as local, fish, all live in these fertile waters.

One open-water community you can readily see, in the right places, is known as submerged

A young brown pelican forages over Pamlico Sound. These birds are commonly seen over North Carolina's estuaries and beaches.

aquatic vegetation. It may be in the freshwater sounds and creeks as well as the brackish and salty ones. Here, larger plants grow beneath the water surface. The aquatic plants look quite different from land plants—they are threadlike or feathery or shaped like bottle brushes. Held by the buoyancy of the water, their stems are soft and move freely with the ripples. Pull them from the water, and they collapse into a handful of green gunk. Drive a motor boat through them and they form a tangled mess around the propeller. But look down on them from a canoe or kayak and they form a mesmerizing, undulating meadow worthy of a fantasy world.

At the opposite extreme from the deepwater and dense marshes are the occasional salt flats, or pannes. These are areas where seawater washes in during the highest tides but then cannot drain away at low tide. As the water evaporates, the salt becomes even more concentrated than in seawater. Sometimes you can see a white crust as the salt crystallizes on the mud. But even if you don't, the sparse and distinctive vegetation tell you that something is different here. Saltgrass, with its distinct, flattened-sideways look, is typical. You may also find the saltworts, their fleshy stems looking more like succulent strings of beads than like typical plants. Bite into one and you will see that they handle the high salt level of their environment by filling themselves with it. Extremely salty places are a rare habitat shared between coastal pannes and desert salt flats. You can find these same plants, or close relatives, around salty springs in Death Valley and other arid places in the western United States.

You can see salt marshes and sounds just about anywhere along North Carolina's coast. The Tideland Trail at Cedar Point in Croatan National Forest offers an easy way to see a salt marsh up close without getting your feet wet, as do the trails at the North Carolina Aquariums at Fort Fisher and Pine Knoll Shores. The ferry and the paddle trail to Hammocks Beach State Park offer a ride through a maze of cordgrass marsh. The largest expanses of brackish, black needle-rush marsh are at Cedar Island National Wildlife Refuge and Swanquarter National Wildlife Refuge. Roanoke Island contains another large expanse, much of it protected by the Wildlife Resources Commission. Submerged aquatic vegetation is harder to find and to get to. It is often located across open water that calls for experience if you are paddling and is often hard to navigate with motorboats. Extensive beds of it are present in upper Currituck Sound, between Knotts Island and Currituck National Wildlife Refuge. For a more sheltered paddling experience, small amounts are present in many tidal creeks, such as those at Goose Creek State Park.

Marsh periwinkle are often visible on the saltmarsh cordgrass at low tide. This salt marsh is at the Rachel Carson Estuarine Research Reserve.

Maritime Grasslands and Beaches

The scene at the final eastward edge of North Carolina is familiar to the many people who live or vacation there. The sweep of the ocean stretching to the horizon, the silent rise and crashing break of the waves, the gently rippled wave-washed sand with its driftwood and shells, are a scene of relaxation on a summer day. The hummocky curves of sand dunes and wind-swept sea oats, with their peculiar flat, dangling flowers, are the backdrop for many a pleasant holiday and many a coastal painting.

If you explore one of the natural coastal islands, you will find other scenes as well. Behind the dunes that line the beach may be the shrub thickets and maritime forests we will talk about in the next chapter. Or you may find flat stretches of sand covered in sparse grass. You may see towering, bare dunes that move with the winds. Or rolling dunes with their sea oats may stretch landward until they meet the marshes and sound. But amid them you may find wet swales nestled, with lush grass and colorful flowers, or tall sedges rising from standing water, or even an open pond.

The sandy barrier islands and spits that line the coast are some of the youngest land in North Carolina, only a few thousand years old or less. They are also some of the least stable. Waves change the shape of the beaches from day to day and from winter to summer. Sand moves with the longshore currents down the coast, building spits, eroding or filling inlets. Ever restless, the wind moves sand up the face of the dunes and tumbles it down the back side. Hurricanes and nor'easters can push the beach back, erode away dunes, and wash seawater and sand all the way across islands. The barrier islands have withstood this pounding not by standing firm but by rolling with the punches. Where storm surges can wash across the island, energy is dissipated, erosion is lessened, and the overwashed sand builds the island landward as the seaward side retreats. On natural beaches, the sand that is washed offshore in storms often is moved back to the beach by the gentler waves that follow. Dunes that are flattened in storms naturally grow back over time.

Few natural communities in North Carolina are as dominated by dramatic physical forces

(above)
You may not see the invertebrates beneath the sand during a walk on the beach, but frantic probing by sanderlings, as they race the surf, hints at the life just below the surface.

(right)
Sea oats at Hammocks Beach State Park trap blowing sand and begin building a new dune.

as those in the maritime zone. Plants that live here bear the full brunt of storms, the risk of erosion of the ground beneath their roots, the potential for flooding by seawater, and possible burial by moving sand. A less visible stress on the plants is bombardment by salt spray. Tiny droplets of seawater, thrown up by waves and concentrated by evaporation, blow on the wind and settle on leaves. Only specialized plants can tolerate the burning salt droplets, and only these plants are able to live at the coast.

Not all of the plants are passive bystanders to these forces however. Sea oats, and its northern counterpart, American beach-grass, actually play a role in shaping the barrier islands. Once established, a grass plant traps blowing sand and accumulates it around its tough leaves. The growth of its underground stems keeps the plant from being buried, and grass and dune grow upward together. This process can be surprisingly rapid. Most of the dunes south of Fort Fisher were completely flattened by Hurricane Fran in 1996. But a well-developed line of dunes reformed within just a couple of years. Once the dunes are built, the grass stabilizes them, reducing sand movement and protecting them from all but the most severe attacks of erosion. Less showy but also quite competent in its own way is the saltmeadow cordgrass. This wiry-leaved grass, a close relative of the cordgrass of the salt marshes, dominates the sandy flats. It tolerates flooding by seawater and quickly grows its way back to the surface if it is buried by overwashed sand.

White-bracted sedge is a showy plant of wet dune swales.

Two distinctive versions of maritime communities deserve special mention, for opposite reasons. The wet dune swales have the lushest vegetation of the maritime grasslands and are the only ones that have many different species of plants. These places are wetlands, and only plants able to tolerate saturated soil without oxygen can survive in them. But, being sheltered from salt spray and wind by surrounding dunes and with fresh water at their feet, these plants enjoy a more benign existence than their more exposed neighbors. This is the place on barrier islands to look for wildflowers. Besides exquisite beauties such as the sea-pinks, the white-bracted sedge is a remarkably showy member of a family known for its modest looks. And one of the most common grasses in these communities, dune hairgrass, sports lovely, lacy purple seed heads that float over the grassland late in the season.

At the opposite extreme are the youngest spits and sand flats, which recent deposition, frequent overwash, and extreme salt spray leave with virtually no plants at all. These unlikely places are choice nesting sites for many of the characteristic birds of the coast. The various species of gulls and terns, black skimmers, and pelicans all make their homes on sand flats, in large colonies with several species

A ghost crab burrow is built into the side of a tiny dune stabilized by sandmat on a sand flat at Hammocks Beach State Park.

together. So do less obvious birds such as the endangered piping plover. These ground-nesting birds seek remote areas, free of disturbance. The lack of vegetation ensures that any predators that might appear can't sneak in unnoticed. With human visitors by the thousands, the barrier islands are not the remote places they once were, and some of the bird colonies have moved to small, unvegetated islands in the sounds, including artificial dredge-spoil deposits. But for the piping plover and for some of the colonies of other birds, the sand flats that are kept free of vehicle traffic and crowds are remnants crucial to their survival.

Maritime grassland communities can be seen at most of the public beaches in North Carolina. Most beach access areas have some grassy dunes, though in some places the dunes and vegetation have been altered by sand fencing or unnaturally dense planting of grasses. You can see extensive natural dunes and overwash flats on Core Banks and Portsmouth Island in Cape Lookout National Seashore, but most of the dunes at Cape Hatteras National Seashore were altered decades ago. Fort Fisher State Recreation Area is a more accessible place to see overwash flats, as well as to see dunes that have reformed after being destroyed by storms. Hammocks Beach State Park has taller, more extensive natural dunes. Jockey's Ridge State Park offers the best example of a large active sand dune. The most extensive wet grasslands are on Ocracoke Island, where they occupy large areas, and near Buxton Woods, where they fill numerous swales. Good examples can also be found in the dune swales on Shackleford Banks and in parts of the Currituck National Estuarine Research Reserve. You can see wetter dune swale marshes and ponds along trails at Buxton Woods Coastal Reserve and the adjacent Buxton Woods part of Cape Hatteras National Seashore.

Wet dune swales like these at Cape Hatteras have the lushest and most diverse vegetation of the maritime grasslands.

Maritime Forests and Thickets

While most coastal scenes are made up of the maritime grassland communities we talked about in the last chapter, sometimes woody vegetation is also part of the picture. Usually it comes in the form of small clumps of shrubs, but in a few places you may see dense forests of short trees. While disturbed examples look ragged, the natural ones often present a striking appearance. Viewed from atop a dune or from the grassy edge, both shrub thickets and forests often show a curved, streamlined surface of solid leaves and twigs that rises smoothly from the ground to the full canopy height, like a hedge pruned by an aeronautical engineer. Sometimes they look like they are crouching behind a dune, not daring to raise a twig above the height of its top. If you explore a large patch of forest, you may find that what looked like a canopy of uniform height is actually quite variable. The tree tops are all at the same level, while the ground rises and falls with the curves of old dunes.

Viewed from the inside, your impression of the forest likely also will be one of dense leaves and odd shapes. The canopy, often not very far overhead, closes you off from the sky, while shrubs may wall you onto the trail. With most of the plants evergreen, this impression changes little from summer to winter. Branches of old live oak trees often snake horizontally as wide as the tree is tall. Even the loblolly pines, usually among the straightest of our trees, tend to be a bit crooked. Some of the plants here, such as flowering dogwood and poison ivy, probably are familiar to you. You may recognize the live oak and the yaupon holly, with its small evergreen leaves, from landscape plantings, though this is virtually the only place in North Carolina they grow naturally. Others, such as Hercules'-club, with massive conical prickles on the trunk, seem to be from another world. What animals you see here are likely to be familiar ones— small birds, perhaps raccoons or deer. But the overall number of animal species in these forests, like the number of plant species, is lower than in most of our natural communities.

These maritime communities are controlled by the same extremes of the environment that we discussed in the last chapter: storms, wind, overwash, and salt spray. However, the trees and shrubs are not quite as tough as the grasses and herbs. Maritime shrub thickets can form

White-tailed deer are one of North Carolina's most widespread animals, occurring in most of the state's natural communities. Here they pause at the border between a maritime grassland and a maritime shrub thicket. The shrub clumps show the characteristic streamlined canopy shape created by wind-borne salt spray.

only with the protection of dunes, and maritime forests can be found only where wider barrier islands and extensive dunes offer substantial shelter.

Salt spray, in particular, has been found to be the major reason why maritime forests and shrub thickets have their distinctive look. The few plant species present are those that can best tolerate salt from above. The thick, evergreen leaves that give these communities their distinctive look are more resistant to salt than other kinds of leaves. The streamlined shapes, too, are caused by salty droplets borne on the wind. They funnel the wind over the canopy and reduce the amount of spray deposited. Any twig that grows above the level of the general canopy gets an extra-heavy dose, and any branch that peeks above the top of the sheltering

Although loblolly pine is extensively planted across the coastal plain and piedmont of North Carolina and often invades disturbed areas, it occurs naturally in the maritime fringe, where it grows along the edges of brackish marshes. Prescribed burning keeps the community in a natural state in this example on the Goose Creek Game Land.

(overleaf)
Live oaks dominate this impressive maritime forest at Springer's Point Nature Preserve on Ocracoke Island, managed by the North Carolina Coastal Land Trust.

Cabbage palm occurs on Bald Head Island, at the northern limit of its native range.

dune is drenched in it. Look closely and you can see the dead tips of twigs pruned back by the salty wind. Even with these adaptations, a storm that kicks up extra-heavy spray often leaves the forests brown for a while.

While the maritime environment is stressful, it does offer some advantages inland places don't have. The spray brings not only damaging salt but also nutrients that feed the plants. Soils as sandy as those of the barrier islands would be extremely infertile inland, but here they can support dense forests. And the fires that once naturally spread across eastern North Carolina and shaped the vegetation were scarcer on the islands. The climate is the most moderate of any in North Carolina. Occasional winters see no frost at all, and summertime high temperatures are lower than farther inland. The distinctive plants of these communities include some that are common to the south but can occur only near the coast at the latitude of North Carolina. This includes the live oak and yaupon holly that are so abundant in maritime forests. Even that tree most symbolic of warmer climates, the cabbage palm, makes its way across the North Carolina border in the maritime forests near Bald Head Island.

While we have talked mostly about the barrier islands, where the maritime environment is most extreme, many of these factors also influence the mainland near the coast. This is particularly true in southeastern North Carolina, where the sounds are small or absent. Maritime forest communities occur in a band a few miles wide in this coastal fringe. There must once have been more of them here than on the barrier islands themselves.

One characteristic of maritime forests that ecologists find interesting is that plant species mix in ways that they don't elsewhere. Most maritime forests occur on the dry, sandy soils of old dunes. But their vegetation includes species of both dry and wet sites. You may remember ironwood from floodplain forests, but here it thrives with sand laurel oak, a tree of dry sites. In the rare version of maritime forest with deciduous trees, southern red oak, a common tree of dry oak forests, is abundant. But so is beech, which normally stays in moist forests.

Maritime forests are rare, but there are a few large protected places where you can see them. Buxton Woods, part of which is a coastal reserve and part of which is in Cape Hatteras National Seashore, is the largest remaining in the state. Trails provide a good look at it. Smaller patches are protected on state lands at Hammocks Beach State Park, Theodore Roosevelt State

Maritime Forests and Thickets

The yellow-rumped warbler is one of the most common birds of the maritime forest in winter. It is the only warbler able to digest the waxes found in bayberry and wax myrtle berries. Its ability to use these fruits allows it to winter farther north than other warblers.

Natural Area, and Bald Head Island State Natural Area, though none are as readily accessible. You can see remnants of what was once one of the largest examples between the houses in Emerald Isle and the other towns of Bogue Banks. The version of maritime forest with deciduous trees is much rarer, but there are protected examples you can visit at the Nature Conservancy's Nags Head Woods Preserve and at Kitty Hawk Woods Coastal Reserve. The mainland maritime forests have almost completely been destroyed by coastal development. A few small remnants can be seen along the roads in the towns of southern Brunswick County, but these are privately owned and not accessible. You can see a more extensive remnant along the road south of Carolina Beach State Park, on buffer lands to the Military Ocean Terminal across the river, but it is closed to the public.

The Future of Natural Communities

We have found tremendous joy and inspiration in studying natural communities and from simply being in them through the years. While no place is fully free from the mark of exploitation and alteration in this long-inhabited land, these are places where you can see beyond it, to glimpse the intricate tapestry not made by human hands. The photographs and descriptions in our tour through the diversity of natural communities represent the best remnants of what nature has to offer in North Carolina. While the pictures in this book were carefully composed to show the natural communities in their best light, sometimes captured in unusual weather or times of day, beauty is always there to be found by those who seek it.

However, natural communities in good condition, where the mark of human hands is faint, have dwindled to a small fraction of North Carolina's landscape. For most of us, natural communities are destinations for special excursions rather than the backdrop of our daily lives. Even some of our most common kinds of communities cover less than 10 percent of the area they once occupied. Some of the less common are down to just a handful of occurrences. Much of the loss occurred in the farming and logging, building and draining of generations past. The successional and altered vegetation that dominates our rural landscapes, and even many of our parks and conservation areas, reminds us that it can take more than a lifetime for nature to recover from such injury.

The loss of what we have left still continues, and too few pieces of our legacy of nature are yet protected. The state's Natural Heritage Program, which inventories remaining natural communities, must constantly update its database to record the destruction of remaining examples. Mines, landfills, crass subdivisions, efficiently managed timber plantations, tastefully planned developments, schools, and ball fields all consume these rare gems. Whether you tally the dozens that are lost across the state each year or witness the plight of a particular favorite place, a sense of urgency and desperation is not misplaced.

These remnants of a healthy and vigorous natural world represent not only a pleasant link to the past but a crucial resource for the future. Their diversity, and the diversity of species and ecological processes within them, represents the building blocks for the ecosystems we

(opposite)
A staff member of the North Carolina Plant Conservation Program replants Venus flytraps recovered from poachers. There is hope that the plants will now be able to survive in this restored wet pine savanna on a Plant Conservation Program preserve.

will depend on in the times to come. We intend these photographs and descriptions to be a record of what is and can be, rather than just an epitaph for what we have lost. In so many of our remaining natural communities, as in the spruce-fir forests, we can see a race between destruction and hope.

But hope remains a contender in this race. Most of the places we have mentioned as opportunities to see particular natural communities are places where they have been protected. Natural communities are protected not only in well-known places such as national and state parks, but also in specially designated areas on national forests, national wildlife refuges, state game lands, plant conservation preserves, coastal reserves, state forests, county and city parks, and a variety of other public lands. There are private preserves as well, from formal preserves of land conservancy organizations to nice spots left alone by individual landowners. There are conservation easements held by conservation organizations, along with the Natural Heritage Program's registry and dedication programs, which allow private landowners and public-land managers to make lasting commitments to protecting their natural areas.

While much of our heritage of natural communities remains threatened throughout the state, there are interested citizens, nonprofit organizations, and government agencies who work to save these places. Some acts are as simple as a landowner learning about what is special on his land and quietly deciding to forgo the profit of logging that area. Some come with outside rewards, such as a landowner donating a conservation easement or offering a bargain sale to her local land conservancy and receiving a tax credit. Some are controversial, such as recent efforts by Raleigh citizens to oppose recreational development in natural areas at the city's Horseshoe Farm Park. Some are complex puzzles with pieces painstakingly assembled over years. For example, the new Chimney Rock State Park has been becoming a reality in the last several years, with much land acquired, more acquisition under way, and dedication of the crucial natural areas in the works. It has involved not only dedicated work by state park staff but also land acquisition and donations by the Nature Conservancy, information from the Natural

Heritage Program, money from the state's Natural Heritage Trust Fund, an initiative by one of the state's best-known family-run tourist attractions, donated private money, political support in the state legislature, encouragement of that political support by citizens, and the cooperation of numerous private landowners. Adding to the ranks and resources of this army of dedicated people could yield a future in which more of our natural areas survive.

Natural-community protection isn't necessarily simple. Sometimes stopping the bulldozers or acquiring land is all that is needed, but usually careful stewardship is crucial to ensure long-term health of natural areas. Natural communities are not artifacts to be put in a museum; they are living entities made of organisms that must be born, live out their lives, and die. These reoccurring assemblages of species vary over time, as weather cycles turn and as the populations in a given place ebb and flow. Processes such as flood, wind, and fire are as much a part of them as the plants and animals.

Perhaps we have introduced you to problems new to you, such as fragmentation of landscapes, invasive species, or loss of natural fire. Some problems, such as the effect of fragmentation, will require protection not only of the best remaining natural community sites, but also conservation and restoration of other areas that connect them. Others, such as prescribed burning and controlling invasive species, may take ongoing effort over many years. Managing protected natural areas depends on dedicated scientists and skilled land stewards. Just as much, it depends on a public that understands the needs and supports the stewards, and on volunteers who are willing to help.

These challenges of stewardship are growing, as the climate warms and sea level continues to rise. We have mentioned specific aspects of these concerns in several chapters, but all of our natural communities will need to adapt themselves in some way. Some will need to migrate inland or up to higher mountains. Others, tied to special environments such as rock outcrops or small basins, will need to evolve in place. Some species may need to move in from farther south, while others that are now common may find themselves increasingly confined to small refuges such as spray cliffs or high-elevation rock outcrops. If we can save our natural communities from development and exploitation, we will need to figure out how to help natural communities evolve and change to survive, while letting them stay natural. Because they are living entities, if we can keep them healthy and diverse, such evolution is possible.

While the difficulties of a changing climate are sometimes regarded as a reason for despair, they could better be viewed as challenges to be met. We already know of some steps we can take to improve the survival of most kinds of natural communities. Many of these actions are the same things we would do to improve their health if the climate were not changing, things such as restoring natural fire, repairing altered hydrology, and reducing fragmentation. Mak-

Scientists and volunteers with the Carolina Vegetation Survey study vegetation in a wet pine savanna. Such research improves our understanding of natural communities and helps us better care for them.

ing them larger, by protecting more area and restoring natural vegetation where it has been degraded, will also improve the prospects of many sites. Natural communities that are healthy, with a diversity of species and processes, are resilient, and the prospects for a vibrant future are good. While nature needs our aid, it also supports us. This race is one we need to win. The ecosystems that will produce the oxygen, filter the water, feed our spirits, and keep our world alive tomorrow will be made from the communities and species that survive today.

What can you do to add to the hope for a future where natural communities and nature in

A trained volunteer helps the Nature Conservancy with a prescribed burn on the Calloway Preserve in the sandhills region. Prescribed burning is helping the community recover from past alterations as well as maintaining natural ecological processes.

general are a part of our lives? You can learn where the most significant remaining examples are and raise your voice for their protection. If you own land where natural communities may remain, you can learn about them and seek advice from experts on managing natural areas for their ecological value. You might wish to pursue a conservation easement or one of the other options we listed above. You can join land conservation and environmental organizations and encourage them to focus on natural communities. You can urge public agencies and representatives to protect sites, to provide the funding needed to acquire them, and, often forgotten, to provide the staff and resources for their management. If you want to put your hands to work, there are many opportunities, from one-time volunteer projects to consuming passions. You can dedicate yourself to detailed study of a particular site or a particular group of organisms, even get the training needed to volunteer for a prescribed-burn crew or invasive-species removal.

What you do when you are not in the woods is important, too. The resources you consume, the waste you produce, all add to the stress on natural communities both far and near. Perhaps most crucially, you can spread your appreciation and concern to those you know. In the end, the best hope for the continued existence of our natural communities is a population of North Carolinians who truly want this heritage of nature to be a part of their home.

Acknowledgments

We would like to offer our thanks and gratitude to all those who helped in making this book a reality. We received many helpful suggestions and improvements from the peer reviewers of the proposal and manuscript. Mark Simpson-Vos of the UNC Press offered sound advice and patiently guided us through much of the process. Laura Cotterman offered editorial aid on portions.

We owe a special debt of gratitude to our wives, Chris Schafale and Leandra Blevins, for their encouragement, help, and patience in putting up with the absence, distraction, mud, ticks, and other hazards that come with being married to an ecologist. Chris Schafale provided substantial advice and editorial assistance on drafts of all text, and Leandra Blevins helped edit photographs and aided with some of the fieldwork.

As is always the case, we owe much to those who helped form our interests, taught us what we know, and encouraged us on our way. David would like to thank Tom Wentworth, whose course on plant community ecology introduced him to natural communities and planted the seeds that became the photographs in this book. Mike would like to thank his parents, Herb and Bennie Schafale, who shared their appreciation for nature, encouraged his own, and provided support for what became a lifetime calling. We would like to thank the North Carolina Natural Heritage Program and its present and past members. Their insights and data provided much of the basis for the natural community concepts and descriptions contained here, and Mike's work for the program provided a substantial part of the experience reflected in the text. Special appreciation is due to Alan Weakley, who has been Mike's greatest partner in community classification, as well as teacher, longtime friend, and facilitator of this book. Thanks also to Julia Fonseca, a longtime friend whose encouragement and belief in the value of this book fed the dream of it over many years. And thanks to all the other friends and colleagues who supported the idea and the effort.

Many people associated with the Carolina Vegetation Survey, the North Carolina Natural Heritage Program, the North Carolina Plant Conservation Program, the North Carolina Botanical Garden, and other conservation organizations helped find locations for photographs, granted access to sites, and shared their perspectives on natural communities. David would like to specifically thank the following people whose expert advice or assistance in the field helped him find images he would never have found on his own: H. Lee Allen, Harrol Blevins, Leandra Blevins, Forbes Boyle, Margit Bucher, Gin Brunssen, Misty Buchanan, Rob Evans, Laura Gadd, Gabrielle Graeter, Donna Hedgepeth, Clay Jackson, Mike Kunz, Mike Norris, Robert Peet, Jesse Pope, Johnny Randall, James Sasser, Stephanie Seymour, Gabriel Taylor, Mike Turner, Alan Weakley, David Welch, Tom Wentworth, Jackie White, Peter White, and Brenda Wichmann.

Finally, we would like to thank the people of the land conservation agencies and land trusts of North Carolina, and all those whose efforts and vision have protected the natural communities shown and described in this book.

Index

Page numbers in italics refer to items pictured in photographs or mentioned in photograph captions.